Gravity Decoded

Gravity Decoded

Exploring the Structure of Space-Energy

Sebastian Borrello

Writer's Showcase
San Jose   New York   Lincoln   Shanghai

Gravity Decoded
Exploring the Structure of Space-Energy

Writer's Showcase
an imprint of iUniverse, Inc.

For information address:
iUniverse, Inc.
5220 S. 16th St., Suite 200
Lincoln, NE 68512
www.iuniverse.com

ISBN: 0-595-20969-6

Printed in the United States of America

To
Bonnie Turgeon Borrello

CONTENTS

LIST OF ILLUSTRATIONS

LIST OF TABLES

ACKNOWLEDGEMENTS

My father was Sam to his customers in the car repair business. He was a skinny guy, able to rebuild a carburetor, smoke a cigarette and carry on a conversation, all at the same time. I spent weekends and summers helping in the garage and dad and I spent evenings building telescopes. He loved precision and developed his own techniques for making nearly perfect telescope mirrors. Dad was an immigrant from Italy and quit school in the seventh grade to get a job and help his family. But he loved learning and instilled in me the desire to know and get as much education as my brain could hold.

In the mid-1950's, when I was a student at Syracuse University, the Physics Department was not the center of campus life. In fact, most students avoided physics as if it were bubonic plague, except for the few of us who reveled in the beauty of its symmetry and the harmony of its logic. I remember well, Nate Ginsburg, his easy-going style of teaching physical optics and the monstrous spectroscopes he built to probe the low energies of hydrocarbon molecules. But it was his office mate, Henry Levinstein, who above all others, taught with flair and humor; he was mentor as well as friend to more students than could fit in a large lecture hall. Henry loved his students, and whether you were a physics major, engineer, or premed student, he treated you with respect and gentle concern. Henry was mentor and father figure, with an encouraging word for anyone in honest struggle with college life. His lectures were punctuated with humor. Each semester he made a quick exit out one door, and jumped in from another door with a white sheet draped over his body, playing the role of a virtual image. In his later years, Henry collected well-designed toys and lectured on "the physics of toys." He did his research on semiconductor infrared detectors and gathered around him a wild bunch of graduate students who became his family. It was an honor to be a "Levinstein student" and most of us kept in contact with each other, as the years went by. Our last big gathering was at Henry's memorial service in Hendricks' Chapel, the summer of 1987. We never forget.

Bill Fredrickson was Chairman of the Physics Department, a job he didn't particularly like, because it stole the time he wanted to teach his first love of physics, astronomy. Dr. Fredrickson, as everybody but his wife called him, asked me to grade papers and operate the university's observatory, with its old refractor telescope, for the public on Monday nights. It was a fun job. Dr. Fredrickson had a little safe in his office from which he made cash loans to students in need of tuition money. I was one of them. No papers were signed; no records were kept; we were simply expected to "pay it back when you can." The staff tried to name the new physics building after him, but even in his eighties, he would have none of that.

Such is life: too many people to thank properly. My mother, Shirley to her friends, was manager of a movie theater during World War II. I got to see a lot of movies. Moviemakers use trick photography, especially now with computer generated graphics, to show action that is physically impossible. But then, moving through solid walls and travel through time are just props for the human drama. At the beach one day, I complained to my mother that someone had smashed my sandcastle. She told me not to worry; someone would always be smashing my sandcastles. Thanks mom, you were right.

Many people have contributed to this book in one way or another. When the space-energy concept popped into my head in late 1996, I told my wife I believed I could develop it into a reasonable theory. She knew it would be yet another time thief. Maybe I can get one of those Hollywood time machines to stretch the clock. Many thanks to Bill

Baldwin, one of those crazy guys who writes science fiction stories, for helping me organize my thoughts and edit this attempt to tell about gravity and to Chuck Cripe for insisting on clarification.

INTRODUCTION

Gravity is a hidden problem. There seems to be too much of it, but nobody complains. Did you ever hear of someone crawling out of a crashed airplane grumble "Damned gravity!" or hear a little old lady complain, "It was gravity's fault I broke my hip." No. Falling is readily accepted as the norm. People have been falling for uncounted centuries and only once in a while has anyone asked, "Why all this falling?" Children ask these kinds of questions, and in their fantasy-play they sometimes pretend that gravity doesn't exist—that flight is effortless and the fall to Earth has little consequence. But soon they learn gravity is real and can exact a heavy price. All forms of transportation and shipping burn a lot of fuel coping with gravity. The builders of bridges struggle with design problems imposed by weight versus strength of steel. The Space Shuttle gorges tons of fuel each minute to reach orbital speed. The infirm are exhausted just climbing a flight of stairs. Is the hold of gravity unshakeable? Always? Can it be nullified or even reversed? Is shielding against gravity or the generation of antigravity possible?

Gravity-free travel or antigravity is like faster-than-light travel in that no one has observed it. Science fiction writers love antigravity because if it were real we humans would be as free as thought. We could escape the bonds of Earth and explore anywhere in the Universe, then come back to tell our more timid friends what a fantastic Cosmos in which we live. Since antigravity is not yet an option feature on cars or rocket ships, the search for the answer to the question of gravity bondage begins with finding out what causes gravity in the first place. We start our search by taking a look at the trailblazers of physical science to find critical clues to this universal mystery, the actual cause of gravity. But at some point it will be necessary to break away from conventional science by looking at the Universe as more than just space and particles. The gravity code is hidden in the fabric of space, and once revealed gravity is simple because the Universe has a simple and elegant design. We will discover how gravity is related to the energy of matter and the energy of space. Then we will begin to understand what we must do to achieve antigravity.

When we attempt to understand the workings of the Universe, we become physicists, all of us, including you, the reader. Nature or *physis*, as the early Greeks called it, can be understood by observation and deduction. We observe a process and use the power of reason to describe the process. As physicists we do not search for absolute truth in the philosophical sense. Rather we construct simple pictures or models that describe what we observe in the natural world.

At the risk of being branded a social outcast, I drifted to the study of physics because I was emphatically told (by physicists of course) only physics shows how matter is structured and how things work. And somehow if you know the how, you know the why. There are shelves of books telling us the Uncertainty Principle, quantum processes, chaos theory, black holes, and so on reveal certain basic truths; most simply hint at the truth of existence. In these books we sometimes find good physics, but never the absolute truth of philosophers or theologians.

Truly, there are beautiful theories of physics that give us considerable insight into the workings of this amazing Universe. In fact, modern physics does such a good job of explaining matter and process that the fringe world of psychic phenomena applies physics to "prove" this or that bizarre idea. Science fiction is more honest if not more totally absurd. But then good science fiction stories are more about human behavior than about science.

A basic problem of learning is that physics professors, like literature professors, parcel out bits and pieces of code from their lofty towers to those empty vessels known as students. This system of teaching actually works pretty well because with authority comes efficiency. Students learn faster when precepts and assumptions are glossed over and seldom questioned. But without the challenge of questioning, knowledge stagnates and eventually petrifies. Knowledge of gravity is a case in point. Gravity ideas froze up three times in history. First with the classical Greeks, second with Newton and third with Einstein.

THE FIRST FREEZE lasted more than 2000 years. Actually, such is modern education that most people still believe as the ancients did. Falling is the norm. Things just belong down, not up. Being down is the natural state of things in of themselves, and so it is natural for things to move to the ground. One is defying the natural order of the Universe by piling stone on stone. This is proven by the fact it takes a lot of work to achieve order and reverse the natural tendency of all things moving toward chaos; witness the average teenager's room. The Greeks codified this thinking of *natural order* some 2500 years ago. And somehow this code became deeply imbedded into Western culture to the point it must not be questioned lest the entire culture comes crashing down. When philosophy, religious or natural, is built on falsehood, it is indeed shaky. Of course, this is easy to see from the threshold of the 21st century. But I dare say many of us would have accepted gladly the learning of that age too, had we lived in it.

Surprisingly, most modern people are still Medieval in their understanding of our world. The average person in the early 21st century thinks intuitively and is usually wrong. But it doesn't matter; people drive their cars, operate computers, use a GPS, even program VCRs without knowing about energy, force, inertia, and so on. In fact, science fiction films violate physical law left and right, yet the average moviegoer doesn't have a clue. Without a doubt, it is perfectly possible to live in both the 21st and 12th centuries at the same time. Engineering makes it possible to live in a high-tech society and really not know how our marvelous machines work. Gaining what understanding of our Universe we do have, has not been easy. It was not until late in the 13th century that the Greek system of logical thinking and natural order finally was freed from its bondage to the ideas of absolutes and perfect movers. The Church in its attempt to rationalize God opened 14th century Church universities to the tools of rational thought. The power of reason was so exhilarating that teachers and students applied it to our physical world as well as spiritual understanding. As the observed world of physical reality was slowly disentangled from spiritual truth, cause as well as effect became rational. The world was ready for a renaissance.

Isaac Newton broke the log jam of Greek Absolutes in the 17th century with his mechanistic description of levers, pulleys and gravity. His genius was the springboard of the Age of Enlightenment that gave birth to engineering and the Industrial Age.

THE SECOND FREEZE: So successful was Newton's understanding of the natural world that Newton's Laws were not to be challenged. By early 19th century the world of science had stagnated again. But with the discoveries of X-rays, atomic decay, and the photoelectric effect, a new science was needed to explain the workings of the atom. Between 1900 and 1930 Newtonian mechanics was transformed to wave mechanics and the theory of mathematical statistics was transformed to quantum statistics to explain atomic properties. Albert Einstein completely removed the vestiges of absolutism buried in Newtonian mechanics and declared that all relationships of space and time are relative. Einstein discovered the curvature of space as the root cause of gravity.

THE THIRD FREEZE. And so here we are, at the forefront of understanding our Universe, immersed in a fantastic muddle of high sounding phrases, intricate particle theory and complex mathematics. Albert Einstein has given us his Theory of General Relativity but the workings of gravity are hidden in its abstract mathematics, and few can break its code. Is General Relativity the final answer? Are we stuck again on another plateau of understanding? Are we permanently frozen in complicated models of how the world works? No, but we do need another way of looking at our Universe to achieve a better, yet simpler model of understanding. I believe that when we get it right even a child will be able to grasp its reality.

Physicists, like children, love to draw pictures to represent ideas and tell stories. Pictures set the scene, but the language may be foreign. The language of physics is primarily mathematics, which is written in shorthand and has specific rules of operation. But there is a big difference between physics and mathematics. Math is so abstract that its rules must be vigorously obeyed to avoid error. When physicists use math it is much easier. The symbols and combinations of symbols have dimension that relate to what they measure, such as size, time, mass, and other dimensions. Keeping track of dimensions is the trade secret of physicists. If a symbol gives speed, it must have the dimension of speed given as distance per unit time such as meters-per-second or miles-per-hour. Equations tell us the relationships of symbols and thus relate dimensions of space, energy, and time. Without dimensions, physics is just a collection of important sounding words with no substance. Keeping the dimensions sorted out leads to understanding and to new discoveries. In this book I use historical observations to construct each equation and reveal when it was just a good guess. Physics is not always trivial, but it is logical. When the path to discovery is very complicated and the results are confusing, chances are the physics is not particularly good. It's best to scrap the confusing stuff, and start over.

In the pages of Part I we will explore the history of gravity. We will watch Galileo Galilei and Johannes Kepler use the power of careful observation and reason to discover the workings of gravity, even though they were still prisoners of Greek mythology. We will find Newton blazing a whole new trail for understanding gravity by inventing a force concept that explains Kepler's laws of planetary motion. Newton himself, however, was greatly disappointed with his new theory because it seemed so irrational. We will see Einstein abolish gravity with his bold concept of curved space, but still be encumbered with

Newton's arbitrary constant. Physicists of the 20th century failed in their search for a grand unification of all physical theory because they saw only particles and not how space connects them.

In Part II we shall combine the great discoveries of history to reveal how gravity is the dynamic balance of matter and space. We will discover that the energy of space is the key to understanding gravity. It is the distribution of this space-energy that defines gravity and gives us its magnitude. Once we break this code, gravity becomes an amazingly simple construct. We will find that gravity depends in a central way on the total matter-energy of the Universe and the size of space. The attraction between any two bodies becomes an interplay of the matter-energy of the bodies and the energy and scale of space. The Universe is no longer arbitrary. Its parts are actively involved in the whole and the Universe as a whole guides the activity of its parts. We will then ask about weightlessness and chart the course for antigravity using our new tools of discovery. Believing that the Universe is simple, we look for a new path and must think anew the workings of this Universe. So open your mind, and let us begin.

Part I

DISCOVERING GRAVITY

What we see depends mainly on what we are looking for

We begin with Galileo Galilei, the first true scientist of recorded history. Unable to completely break with Greek mythology, he nevertheless established the principle of quantitative measurement and the deduction process of theory. About the same time, Johannes Kepler, a prisoner of mysticism, deduced his theory of planetary motion by being absolutely faithful to the precise measurements of planetary orbits made by Tycho Brahe. With Galileo and Kepler setting the stage for enlightenment, Isaac Newton broke the chains of Greek convention by claiming that gravity is a property of matter and that bodies of matter pull on one another. When Newton developed his law of gravity in 1667, he marveled at its simplicity, but was greatly bothered that it was "action at distance," an action he did not understand. The sun pulls on planets, and Earth pulls on us, but there are no strings attached. Gravity acts at a distance. Albert Einstein in 1912 abolished gravity by showing that curved space can account for the appearance of attraction. But Einstein does not tell us how matter curves space. Both theories are great triumphs of science, even though they still leave us hanging on the why of it all. Einstein knew that the source of gravity is a property of space, and he did his analysis using a special kind of geometry. But his

1

approach is very difficult to follow. Now, liberated from the bonds of convention, we will set a new course of discovery that will lead to new ideas about our Universe.

CHAPTER 1

Falling

For nearly twenty centuries, Western schools taught from a basis of Greek philosophy. The natural world, it was believed, presented too many illusions to those who would learn by observation. The only reliable way to understand nature was by a pure intellectual process called induction. Ideas are formulated, and if an observation doesn't fit, the observed must be an illusion. Falling followed this pattern.

With the Greek concept of induction, all things tend to move to the ground because being down is most natural. Smoke may rise, but it will eventually fall. And since the natural world is not pure like thought, measurements of falling would be misleading.

So far as we know measurements of falling were not conducted and reported until the late 16th century. And it was then that science began. Galileo Galilei while teaching mathematics at the Universities of Pisa and Padua from 1590 to 1600 made a measurement and thought about it. Like others of his day, 25-year-old Galileo could see that when a cart broke free of its hitch on a hilly street it moved slowly at first and steadily picked up speed going down hill. The thing that intrigued this Renaissance man was not the descent down the hill, for rolling downhill is natural after all. His focus was how the descent was taking place. When Galileo dropped objects off his balcony, he suspected that the speed increased during the fall, but the fall happened too fast to be sure.

In a flash of brilliance, he guessed that the cart rolling down the hill was a slow-motion version of vertical falling. Scientific breakthroughs do occur as flashes of brilliance, but only after many hours of concentrated thinking. The mind must be prepared for the flash. This is true for any endeavor of the mind. Of course Galileo was right; the rolling cart was a controlled fall, controlled by the slope of the street.

Now the science begins; Galileo goes to his workshop and builds a smooth pair of rails and little carts with smooth metal wheels. He tilts the rails to form a shallow incline and lets the cart roll down the ramp as shown in Figure 1. As the cart rolls, he marks its position on the rail with a little paintbrush with each tick of his noisy pendulum clock. After a lot of practice he gets good at synchronizing with the clock. He measures the distance traveled with each tick. The distance traveled between ticks increases. And since speed is the distance traveled per unit of time, the speed is in fact increasing as the cart descends.

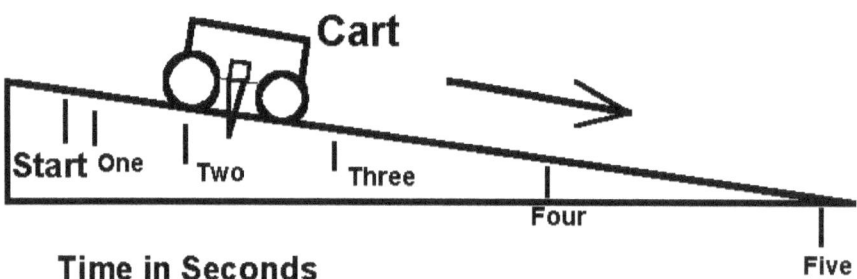

Figure 1. As the cart rolled down the shallow incline, Galileo marked the progress as his clock ticked off the seconds. When he elevated his 200-inch-long ramp 7 inches high, the cart moved 6.7, 27, 60 and 107 inches from START at the marked times.

Since the clock ticks are always the same, tick to tick, he observes speed is in proportion to the distance between ticks. As a teacher of mathematics it is quite natural for Galileo to make a chart showing the progress of the cart to help him visualize the relationship between distance and time.

The difference in distance between pairs of ticks measures the increase-in-speed we now call acceleration. Galileo calculates average speed for each interval of time by taking the ratio of interval distance to interval time as he shows in Figure 2. The average speeds for the 7-inch-high ramp are 6.7 inches per second, 20.3 in/sec, 33 in/sec, and 47 in/sec in the four intervals. Now Galileo goes one step further and charts the speed in relation to time.

The speed is increasing, but low and behold the rate of increase of speed is constant as he illustrates in Figure 3. Or nearly constant. His data have experimental error, and he knows it, but it is good enough to show that falling is well behaved. He is amazed. He changes the angle of the ramp, and puts a variety of materials in the cart such as rocks or lead. The rate at which speed changes is the same, no matter where the cart is on the rails, and no matter what is in the cart. Then the revelation hits, "all falling objects speed up the same." Science is born. Thoughts are tested by experiment, and a new concept is derived.

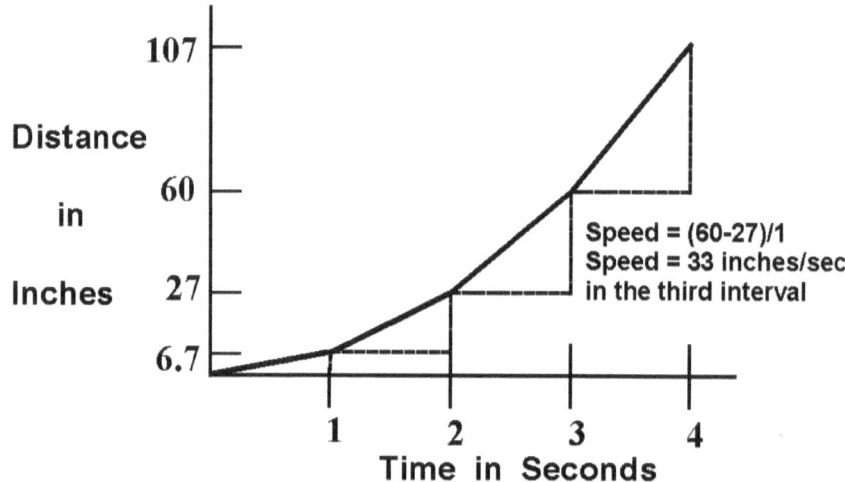

Figure 2. Galileo charted the distance traveled each second. Speed in each interval of time is the interval distance per unit of time.

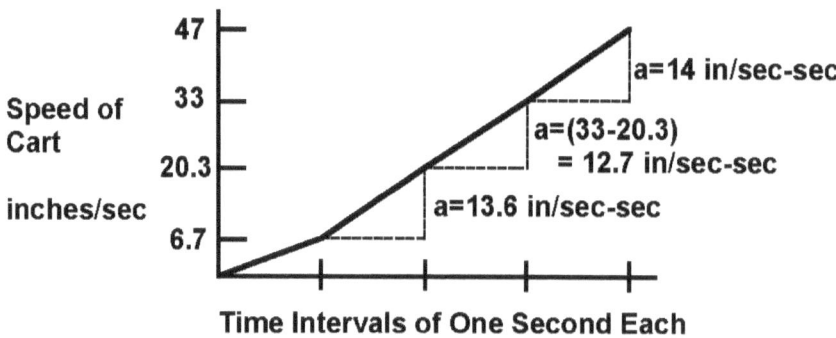

Figure 3. Galileo charted the speed for each interval of time. He calculated the change in speed for each measured interval of time. This change in speed, acceleration, was nearly the same for each interval. The acceleration of falling has a constant value.

Galileo observed that acceleration varied as the height of the ramp, and being a teacher of mathematics, figured correctly that his results for a ramp at low angle could be extrapolated to vertical falling by simply multiplying by the ramp's length to height ratio. He multiplied his measurement of 13.4 inches per sec- sec by the 200 over 7 ratio and got 383 inches per sec-sec. Dividing by 12 inches per foot, Galileo got 31.9 feet per sec-sec for the value of the acceleration for vertical falling. No wonder free falling was difficult to measure; speed increased nearly 32 feet per second each second.

Galileo measured distance using the old Roman unit of distance, *uncia* (sounds like ooncha), which was one twelfth of a foot. Our words inch and ounce come from this Latin word. Galileo's feet and inches were not much different from what Americans use. Later in this book, we will switch to the metric system, because its conversions are easier to follow.

Galileo noted that when the elapsed time from the start of motion, denoted by the letter t, was used, the distance of falling must be written:

$$s = \tfrac{1}{2}\, a\, t^2 \qquad\qquad (1)$$

This is our first equation! It says that when starting from a stop, the distance the cart travels, given by s, is equal to half the measured acceleration, a, times the square of the elapsed time, t. As you will remember from high school, we write our equations in the modern way, such that terms next to each other implies that they are to be multiplied. Addition will always be made explicit by the plus sign (+) and subtraction will always be denoted by the minus sign (-). Division is denoted by the divisor line (__) or slant sign (/).

Let's try Equation 1. If the free fall part of a bungee jump lasts for 2.2 seconds, the flyer will fall 1/2 times 32 times $(2.2)^2 = 77.44$ feet before the fall is slowed by the elastic cord. Let's check the dimensions.

$$\text{DISTANCE} = \frac{\text{FEET}}{(\text{SECOND SECOND})}(\text{SECOND SECOND}) = \text{FEET}$$

The dimension of seconds times seconds appears above and below the divisor line. They cancel each other, leaving only the dimension of feet. The physicist always checks dimensions to be sure the equation is dimensionally correct.

Of course science is more than just measuring. The results must have a meaning that extends beyond any particular test. The struggle is to find that meaning, which we call physical theory. Galileo discovered that the distance covered by the cart increased as the square of time. His theory became:

> *All falling objects speed up or accelerate the same, with the speed increasing proportional to time and distance increasing proportional to the square of time.*

His new theory passed test after test, balls or carts rolling down ramps, balls dropped from towers, and so on. Although Galileo learned how falling progressed, he still reasoned that falling was natural for all objects. He had discovered that all objects fell the same, *not why*. And by that time of his young life, he had made his first telescope and was on a new path to great discoveries, as well as the trouble they brought him. But his little carts delivered yet another revelation.

When the carts ran off the ramp they would continue to roll across the table. Galileo extended the table. He noticed that the cart would stop on the extension which was left rough by the carpenter. But when the table was finished smooth, the cart kept rolling on. The Greeks taught that all motion needed a "mover" to keep the object moving. Ships needed oars or sails. Chariots and wagons needed horses. Birds flapped their wings. Galileo questioned this Greek thinking. He discovered that as wheels and tables were made smoother, the carts rolled for a greater distance. By marking the progress across the table by pen and clock he found that the speed tended to be constant as polishing reduced the friction. Galileo declared that *an object set in motion on a smooth level surface will remain in motion at uniform speed unless something slows it down.* No mover is needed to maintain speed except to overcome friction. When driving on a level road, the gasoline we burn to maintain constant speed gives the car just enough power to overcome wheel friction and air resistance. Galileo taught constant acceleration is a fundamental property of falling and moving objects have unchanging speed when not impeded.

Although all carts zipped along the table at the same speed for any given angle of the ramp, Galileo noticed that the heavier carts were more difficult to stop. This was an ancient idea but Galileo and those who followed quantified this property they called *impetus* and we call *momentum*. Momentum is related to the product of weight times speed. More on that later. The stage was set for Newton, almost.

CHAPTER 2

Heavenly Harmony

As the 17th century began, another thinker, Johannes Kepler was busy, extremely busy, studying the mountain of planetary data accumulated by astronomer Tycho Brahe. From 1576 to 1597, Brahe kept track of a planet's location by referencing it to the background of stars, which have almost no apparent motion relative to each other. Each clear night he would measure the angular progress of whatever planets were visible using his quadrant, a sighting device like a giant protractor. Brahe used Earth's rotation and path around the sun, being well ordered events, as a celestial clock to measure time. Kepler was a brilliant mathematician for his time, but his science was flawed by his purpose. He was determined to show that the planetary orbits were arranged by God to fit the geometrical patterns of "perfect" polygons.

Ancient peoples long regarded the night sky to be a revelation of the gods. The positions of stars and the motion of planets were code for events, past and future. Nothing was by chance. Read the code and control your destiny. The Greeks observed that the structure of a musical instrument determined its musical scale. To them the structure of the heavens followed some kind of musical harmony, "the music of the spheres." Kepler too, believed the planets were arranged to achieve a mystic harmony. In his pursuit, he was guided by the heliocentric theory of Polish astronomer Nicolaus Copernicus who placed the sun at the center of the Universe, but insisted that the planets moved in perfect circles about the sun. No matter; Kepler's basic honesty, determination, and precision forced him to discover that the planets moved in elliptical orbits, not perfect circles.

As he generated each orbit by plotting Brahe's data on paper, he placed the sun at one foci of the ellipse and constructed small angle triangles out to points on the orbit designating equal intervals of time. It was difficult and tedious work because Kepler made complete use of the preciseness of Brahe's data. The nearness to a circular orbit (1-e, where e is the eccentricity) is given as a percentage in the last column of Table 1. Venus and Earth have nearly circular orbits, but Mercury's is quite elliptical. A pattern was showing up. The angles were bigger as the planet approached its closest point to the sun, called *perihelion*, as indicated in Figure 4. He calculated the areas of each sector and discovered that the areas swept out in equal times were equal. Some kind of order did exist. This motion required that the planets speed up as they approach perihelion and slow down as they return to *aphelion* (away from helios, Greek for sun). Distance and speed were somehow related.

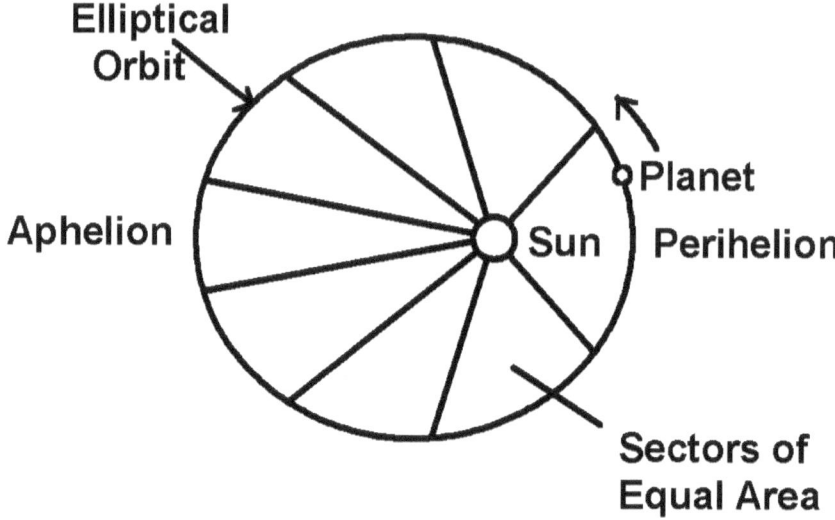

Figure 4. Kepler used Brahe's data of planetary motion, to discover that planets move in elliptical orbits. His calculations showed that when the ellipse is divided in sectors of equal travel time, each sector has the same area. This proved the planets speed up, as they move from aphelion to perihelion.

Kepler reasoned that the orbital mechanics of a planet suggested a mathematical relationship existed between a Planet's distance from the sun and its orbital speed. He abandoned his quest of perfect polygons and searched Brahe's planetary data for clues to planet mechanics, as well as, he hoped, a mathematical harmony. He pressed on.

Kepler let his love of mathematics and Brahe's data guide him in his quest for planetary rhythms. He listed for each known planet its average distance from the sun and its period of revolution as given in Table 1 (here we use miles and seconds, but any units of distance and time would work). Assuming that planets obey some simple power law, he tried different powers of distance and time looking for a pattern (some of his possible trials are shown in the table). And sure enough, the ratio of the cube of the distance to the square of the orbital period was nearly identical for all the planets. Kepler was elated;

From these exacting studies, Kepler gave us his three laws of planetary motion:

Planets move in elliptical orbits.

The speed varies such that the area of each equal time segment, is equal.

The cube of a planet's average distance, r, to the sun divided by the square of its orbital period, T, is the same for each planet.

The fact that planets, moons and satellites have elliptical, rather than circular orbits is not a mystery. Both the ellipse and the circle are sections of the cone. Take an ice cream cone and place it upside down on a table. Any cut that passes through the cone leaving a continuous cut edge of cone has generated an ellipse. Ink the exposed cut and press it to paper to make a nice elliptical print. When the knife passes perfectly parallel to the table the cut edge is a circle. The circle is *the* special case of the ellipse. The circular orbit is *the* special case of orbits. It is very rare. When a chunk of space debris, coming into our solar system, has insufficient speed to escape, it will orbit the sun, and the orbit will most likely be an ellipse. The Space Shuttle has a nearly circular orbit, and it does take a rocket scientist to make it so.

PLANET	DISTANCE FROM SUN, R IN MILLION MILES	PERIOD OF REVOLUTION, T TIME IN MILLION SECONDS	R/T	R^2/T	R^2/T^2	R^3/T^2	SIMILARITY TO CIRCULAR ORBIT 100% FOR A CIRCLE
Mercury	35.98	7.61 (88 days)	4.73	170	22.35	804.0	79.4
Venus	67.2	19.41 (224 days)	3.46	233	11.99	805.5	99.3
Earth	92.96	31.56 (1 year)	2.95	273.8	8.676	806.5	98.3
Mars	141.67	59.36 (687 days)	2.39	338	5.696	806.9	90.7
Jupiter	483.7	374.3 (11.86 yrs)	1.29	625	1.670	807.8	95.2
Saturn	886.7	929.8 (29.46 yrs)	.954	846	.9094	806.4	94.4

Table 1. Planetary data and ratios of distance to orbital period are listed. Kepler discovered the unique ratio r^3/T^2 was common to all planets. It was beautifully simple, but it did not reveal to Kepler which harmonic scale it was related to. Kepler did not find his answer because there was none.

All this by 1609. It was beautiful, but no one could figure out why the planets behave this way. Kepler kept looking for some kind of geometrical harmony as the root cause, as if God moved the planets in tune with unknown celestial perfection. But alas the world would have to wait another 57 years for Newton to break the code.

CHAPTER 3

Falling by Attraction

We can give thanks to Galileo and Kepler for finding the mechanics of falling and the pathways of the planets. They did it the empirical way, by combining experimental evidence and mathematical modeling. Falling objects and planets were reliably described with mathematical rigor. The Universe seemed well behaved. Did the gods of ancient Greece settle down to obey some kind of exact order? Does God constrain falling and the motion of planets to satisfy his celestial math? Or does matter itself dictate what it does? Newton found an answer by changing the rules of falling.

Isaac Newton was a student at Cambridge University when the bubonic plague broke out in England in 1666. He decided to go home until the disease ran its course, so the story goes. I however, think Newton was beginning to see a connection between falling and planetary motion and wanted private time to work it out.

Picture Newton on his mother's farm sitting by an apple tree. He sees a neighbor's horse in the next pasture and, knowing how much horses love apples, he tosses one over the stone fence. It flies up and then falls down. Newton thinks about it. He throws another apple. He asks himself, when does the apple start to fall? I give it an impetus; it goes to some height, and then falls to earth. If Galileo is right, it should keep on going in the line my hand directed it until something resists its motion. But no, it is leaving that line of directed flight as soon as it leaves my hand. By God, it must be falling all the time it is in the air, even when it is going up. Falling is occurring during the whole motion! It must be that, or else it would stay on a straight path. The path is a curved path all the way. But why is it falling all the time? It is as if something unseen is pulling it down. Something is interfering with its natural motion.

Notice how Newton at age 24 had turned the accepted idea that falling is due to something inherent in the object to the new idea that falling is caused by something else—

something unseen. This is an amazing breakthrough in human thinking. Newton uses a simple observation and applies to it Galileo's discovery that objects retain a uniform, straight-line, constant-speed motion, unless acted on by an explicit influence, a push or a pull. A force. And this force must have something to do with Earth, because all things hurled upward come down. Since Galileo proved all such objects accelerate the same, Newton concluded that some force of Earth causes this specific acceleration. The acceleration is proportional to the strength of the force. Newton may have gotten this far before 1666 and needed time away from the distractions of college life to work out his theory of attraction.

He made another discovery, almost by accident. When Newton pushed on a parked wagon, no motion occurred until the push force reached a value peculiar to the given problem—say 88 pounds of force for the given situation. Up to a push-force of 88 pounds, one can say the wagon was pushing back with equal force. When motion began, one might say Newton had begun pushing harder than the wagon was pushing back.

But Newton noticed that less pushing force was required, once the wagon got moving. He concluded that no matter what the motion—

standing still, constant speed, or acceleration—the wagon always pushed back with equal force. The force of reaction always equaled the force of action. Every action had its equal reaction.

Pondering the ideas of mutual attraction, acceleration, and the equality of reaction, Newton discovered the underlying principles of planetary motion. How he did this is not really known. But he most certainly knew that whatever the law of attraction was, it must be consistent with the results of Galileo and Kepler. It is believed he considered the moon to be like the apple. The moon is moving through space around Earth, and if Earth suddenly lost its influence over the moon, it would fly off in a straight path to some unknown corner of the galaxy. A similar result is obtained when a rock tied to a string is spun around and released. The rock goes off in the direction it was traveling, at the moment of release. But Earth holds fast, and our moon follows a nearly circular path around Earth. Likewise, Earth, in its path around the sun falls to the sun just enough to maintain its steady orbit.

When he returned to Cambridge, Newton declared his theory of planetary motion and was able to derive mathematically Kepler's relationship between distance and orbital period. His declarations were something like the following.

> *Acceleration of a body of matter is proportional to the acting force. When a force accelerates matter, the degree of acceleration depends inversely on the amount of matter influenced by the force.*

Newton was thinking in terms of what we call *mechanics*, which is the study of force applied to rigid bodies. We experience force as a push or pull. A rigid body is a chunk of matter that does not change in size or shape when it is pushed or pulled. There are no exactly rigid bodies, but things like tables, bricks, and forks are rigid enough to demonstrate the effect of force.

Newton's big problem was quantifying matter in such a way as to achieve a law of acceleration that would be *universal*, something common to all bodies of matter. He settled on using the property of weight to identify the relative value of matter from body to body. But he wanted an absolute measure of the innate character of what reacted to force. The obscure nature of his writings on this subject indicates he was not successful with any definition of matter.

We call the property of matter that reacts to force *mass*. Denser substances of the same volume have more mass. An iron ball has more mass than a wooden ball of the same size.

Just what is mass? From Newton's law of acceleration, it is what resists force; it results in inertia, the tendency for a stationary body of matter to stay at rest—or if in motion, it is the tendency to stay in motion and resist change of that motion. Mass and inertia are related. Does this help define mass? Perhaps, but a standard of mass is needed, such that all measurements of force have a common reference. And so by the late 19th century, physicists finally agreed on *the standard*. The Standard Mass is a chunk of platinum, kept at the International Bureau of Weights and Measures near Paris. By *definition* it is exactly one kilogram of mass. Put this one-kilogram of mass on an American weight scale, and gravity at the Earth's surface will pull on it to register 2.2 pounds of force. The National Institute of Standards and Technology near Washington, DC, keeps a copy of this standard mass, as a secondary standard (more than two hundred years after Newton, Einstein gave mass a deeper meaning, as we shall discuss below).

Some physicists have proposed that a chunk of silicon be used as the Standard Mass because silicon can be purified to less than one foreign atom in 10 billion. And pure silicon is quite stable. More importantly, silicon has a well-defined crystal structure, and eventually the Standard Mass may be defined as a specific number of atoms of silicon. Then, anyone who can count silicon atoms in a chunk of purified silicon by using X-ray or particle deflection techniques may produce a secondary standard that meets the need for the most precise measurement.

We will now discuss Newton's results in the simple language of mathematical equations rather than his method of complicated geometrical patterns. In fact, Newton did not use algebraic notation to represent physical quantities, making it very difficult to follow his reasoning. So we use simple algebra.

Newton's law of acceleration caused by a force, F, may be written:

$$\textit{Acceleration Law: } a = F/m \qquad (2)$$

Mass, given by the letter m, has no more meaning of action than given by this equation or the standard chunk of platinum located near Paris. But even though mass is arbitrarily quantified, Newton's Acceleration Law is extremely important and useful. We may read this equation in different ways. This equation says several important things. When there is acceleration, a force is acting. When more matter is being accelerated, the force must be greater to achieve the same acceleration. And when we rewrite Equation 2 to say F = ma, we may calculate the force required to accelerate a given mass to a certain value of acceleration. Equation 2 is universal. It works everywhere and for all mechanical forces. Galileo found that the force of Earth's attraction (even though he didn't think in these terms) produces an acceleration of 32 feet per second for each elapsed second. Galileo gave us the value of acceleration produced by Earth at its surface. Newton said, the source of acceleration is the mutual pull between Earth and body which produces a force of so many pounds for a given mass.

Let's see how this works. My car weighs 2,600 pounds. Using Newton's law of acceleration, it has a mass of m = F/a = 2600/32 = 81.25 units of mass (in our English system of units, mass is given in *slugs*, no kidding). Or try this, using Galileo's concepts. I want to reach a speed of 60 miles per hour (88 feet per second), in 8 seconds from a dead stop. Keeping my acceleration constant, with careful control of the accelerator pedal, I must accelerate at 88/8 (speed divided by elapsed time) = 11 feet/sec-sec. And since F = ma, the engine and drive train must produce a force of F = ma = 81.25 times 11 = 893.75 pounds of force (also called thrust). This tells me what engine horsepower I will need (auto engineers can convert thrust requirement to horsepower needed).

In the metric system of measurement, mass is measured in kilograms, force in newtons, and acceleration in meters per second-second. The metric system was established to provide a universal system of weights and measures; it is based on factors of ten, which is a decimal system.

CHAPTER 4

Making a Good Guess

Newton's most astounding declaration upon his return to Cambridge was, *there is a force of attraction, a force of gravitation, between all bodies of matter,*(rocks, celestial spheres, balls of cotton, atoms, literally all bodies of matter attract each other).

How Newton came to his law of gravity is not clear. In fact it is a complete muddle. But then, that is often the case when one is exploring for new ideas. It is only after the theory survives the tests of experiments and becomes immersed in a larger body of logic that simpler means of derivation are found. In 1666, Newton had as a starting point the results of Galileo and Kepler and his own concepts of force. These were:

- All falling bodies have the same acceleration. Galileo.
- A body displays uniform motion (constant speed or at rest) unless a force is applied. Galileo and Newton.
- The planets move around the sun, such that the ratio of the cube of the distance to the square of the orbital period is a constant (this ratio has the same value for all planets in our solar system). Kepler.
- Acceleration is caused by force, and is proportional to the magnitude of the force. That is: $a = F/m$. Newton.
- Every action has an equal reaction. Every force is balanced by another force. Newton.

From this base of knowledge, Newton declared his Law of Gravity:

For any two bodies, this force of mutual attraction is proportional to the product of their masses divided by the square of the distance, r, between the centers of each body.

The stage is now set to see Newton solving the problem of planetary motion. How did he do it? Unfortunately, we do not know. Newton did not reveal his path of discovery. As is often the case in physical science, once a discovery has been made and experimentally verified, the history of its original development disappears. Before we try to reconstruct Newton's path of discovery, let's see what his Law of Gravity says.

We use the shorthand language of mathematics. Subscripts help identify the force of gravity, F_G, the mass of body 1, M_1, the mass of body 2, M_2, and r^2, the square of the distance between the centers of body 1 and body 2. We write Newton's famous law of gravitation,

$$F_G \propto M_1 M_2 / r^2 \qquad (3)$$

This is not an equation; there is no equal sign. It is written as a proportionality, with the symbol, $\propto$. Newton wrote it this way because he was not sure if the force of gravity depended on any other physical quantity. Even so, there is one very important problem to be sorted out. The right hand side of Expression 3 does not give force. It is not dimensionally correct. If we write out the expression in its dimensions, we read it as follows,

kilogram kilogram
meter meter

These are not units of force. Newton's gravitation law must be changed to make it work as an equation and give the dimension of force. The dimensions must work out, or else it is not a useful law. The fix is to put in a multiplying factor that yields the correct unit of force. It must have the dimensions,

Meter meter newtons
Kilogram kilogram

This factor is called the *constant of gravitation*, and is designated by the letter, G. By the mid 18th century, physicists had modified Newton's discovery to give us a gravity *equation*.

NEWTON'S GRAVITY EQUATION:

$$F_G = G \, (\, M_1 M_2 \, / \, r^2) \qquad\qquad (4)$$

Let's check the dimensions. The first term is the dimension of G and the second term is from $M_1 M_2 / r^2$:

Meter meter newtons kilogram kilogram
Kilogram kilogram meter meter

From the rules of algebra, all the meters and kilograms cancel out leaving just newtons, the unit of force. If this seems contrived, you are right. G is a contrivance to make Equation 4 work. And in fact, it does work very well, provided the value of G is correctly obtained by experiment.

Physicists use constants to change proportionalities to equations. Constants of proportionality are tools used to construct equations. It is not a trick in the sense of being devious because the physicist goes to great lengths to experimentally measure the value of the constants he or she uses. For example, let's say you discover that the number of people that show up for a meeting is proportional to the number of cookies provided. Double the number of cookies and twice as many people come. If you provide 100 cookies, how many will come? We don't know until we know the constant of proportionality. So we try an experiment. We set out 20 cookies and 10 people show up. We test it again and set out 40 cookies. Twenty people show up. The constant of proportionality is 10/20 = 20/40 = 0.5 persons per cookie. Clearly, 100 cookies will result in 0.5 times 100 = 50 people. Knowing the value of the constant gives us the power of prediction.

The gravity force is the attraction between any two bodies with their individual centers of mass a distance, r, apart. In the American measuring system, force is measured in pounds, and in the metric system it is measured in newtons, as an honor to Sir Isaac Newton's contributions to science. Newton's result is elegant in its mathematical simplicity and powerful in its physical meaning, as we shall see.

Newton, however was not particularly happy with his great discovery. The need for the constant G meant that it must be determined by measurement before he could evaluate planetary orbits. And more than this, his law of gravitation was not independent of arbitrary definition. Because to find G one must define mass and set up an experiment to measure G. Rewriting Equation 4 to obtain G, gives,

$$G = F_G\, r^2\, /\, M_1\, M_2 \quad (5)$$

During the 18th century, several physicists measured G by hanging iron or lead balls on long strands of fine wire, and measured the sway of the wire as the two massive balls pulled on each other. Henry Cavendish refined the experiment, and in 1796 measured G = 6.670 E-11 newton-meter squared/kilogram squared (nt. m²/kg²). The E-11 term is the abbreviation of exponent of 10 raised to the −11 power. This way to write exponents is now common with the language of computers. G is a very small constant, meaning that it takes a lot of mass (collection of atoms) to have a measurable force. G may also be expressed in American units; G = 3.44 E-8 pounds—feet squared/slug squared (remember, a slug is a unit of mass).

This problem of an arbitrary definition of mass and the artifact G haunted Newton. His law of gravitation is simple and works very well, but it lacks fundamental elegance because of the arbitrariness of mass and G. Much later, Albert Einstein overcame the difficulties of mass, but his field theory still required the use of G to produce a meaningful measure of force. Although Einstein abolished gravitational force, it does not go away. As we shall see later, elegance, free of arbitrariness, can be achieved. Mass is energy, and the Universe tells us what G is. When we finally reach elegant simplicity in Part II, we will know what gravity is. But let us first follow Newton's lead a little farther.

His lectures made him famous and earned him Knighthood, but Sir Isaac Newton waited more than 20 years to write the compendium he called *Mathematical Principles of Natural Philosophy*. The results are brilliant, but the geometrical methods he presents, as a derivation, are extremely tedious, and very difficult to understand. It is likely that Newton searched diligently for a better theory because in 1691 he wrote:

"That gravity should be innate, inherent and essential to Matter, so that one Body may act upon another at a Distance thro' a Vacuum, without the mediation of anything else, by and through which their Action and Force may be conveyed one to another, is to me so great an Absurdity that I believe no (competent thinking) Man…can ever fall into it."

For Newton, "a most subtle spirit" maintained this "action at a distance". This subtle spirit, we now know to be space-energy. But that is getting ahead of this story of discovery. We look next at how the law of gravity is confirmed by Kepler's law of planetary motion, which was empirically derived from the observational data of Tycho Brahe.

CHAPTER 5

The Moon is Falling

If we assume that Newton plucked his law of gravity out of thin air, it would be necessary to test it by measurement, or at least show how it could prove Kepler's law of planetary motion. The following is what physicists call a "thought experiment" that uses mathematical and physical logic to go from one concept to another to check for contradictions. Good science yields laws that are testable, and can always be tested for contradictions. A theory is not valid until it is non-verifiable, which means it may be deemed incorrect if it can not be subjected to verification tests. Science is always testable, and if a theory fails key tests, it is rejected.

Newton had the insight that if he could throw an apple fast enough, it would travel around the entire world before it hit the ground (neglecting the friction of air, of course). And so it was a natural extension of this thinking that he reasoned the Moon keeps its path around Earth by falling just enough to compensate for its natural tendency to fly off into space, as a stone would leave a sling when the pouch holding it is released. Considering the Moon's path under the condition that gravity was to suddenly cease its grip, Newton was able to establish the precise form of Kepler's theory.

In Figure 5, the Moon suddenly leaves its orbit at position P, goes a distance d, and then falls back toward Earth along h, as gravity is turned back on. This scenario is quite artificial, and should be considered only in the limit, where h is very much smaller than the distance to the Moon. Newton thought of the Moon as continually falling, in order to stay in its orbit.

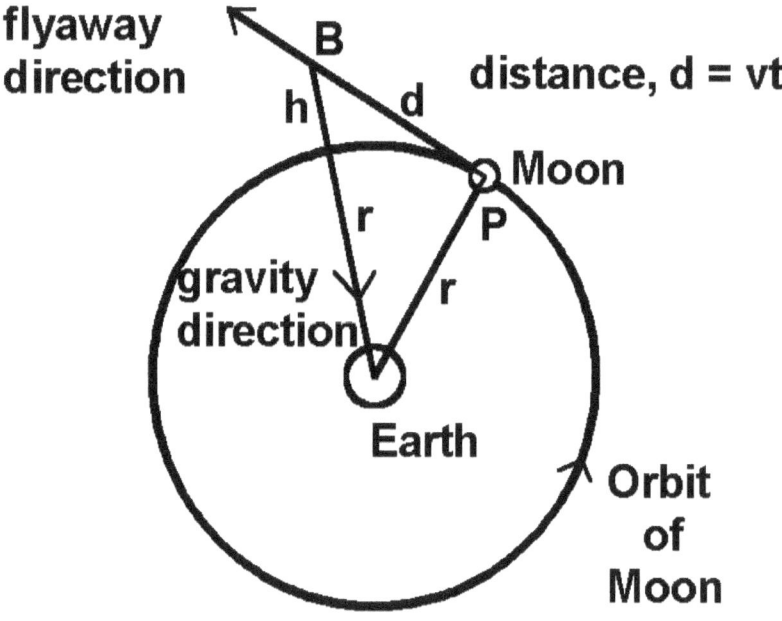

Figure 5. Isaac Newton thought of the Moon as falling under the influence of gravity. Moon would travel at its orbital speed, v, from P to B, if gravity was suddenly turned off during the time it would take to go a distance, d. Newton used this imaginative scenario to prove Kepler's theory of planetary orbits.

We will use algebra and the dimensions in Figure 5, to show how Newton's law of gravity, leads directly to Kepler's law of planetary motion.

The right triangle given by Earth, Moon, and position B, may be solved using the Pythagorean theorem (the hypotenuse squared equals the sum of the squares of the other two sides, remember?). The side h+r is the hypotenuse.

$$r^2 + d^2 = (h + r)^2 \qquad (6)$$

The right side of Equation 6 may be written (h+r) times (h+r), which is multiplied out to give:

$$r^2 + d^2 = r^2 + h^2 + 2rh \qquad (7)$$

We can simplify this equation by throwing out any very small, additive terms. The term h^2 is an added term, and compared to 2rh, we must have h^2 much less than 2rh, or equivalently, h much less than 2r. The distance to the moon is 240,000 miles, and h may be as small as we like in this thought experiment. We throw out h^2 with no loss of accuracy. The r^2 terms cancel each other. This equation now simplifies to:

$$d^2 = 2rh \qquad (8)$$

Next, we evaluate the dynamics of d and h. The orbital speed, v, of the Moon is just the circumference, $2\pi r$, of the orbit divided by the period T, the time for one orbit, or $v = 2\pi r/T$. The Moon is traveling at speed v in its orbit, and when gravity quits Moon goes off tangentially with its orbital speed, v. The speed is also just the distance, d, traveled in a time t. That is, $v = d/t$. Equating these two expressions for v, we have:

$$\frac{d}{t} = \frac{2\pi\ r}{T} \quad \textit{and squaring,}$$

$$\frac{d^2}{t^2} = \frac{(2\ \pi\ r)^2}{T^2} \tag{9}$$

Solving for d^2, we get:

$$d^2 = \frac{(2\pi\ r)^2\ t^2}{T^2} \tag{10}$$

The distance of falling, h is given by Galileo's discovery that $h = at^2/2$ (using Equation 1) for acceleration, a. Using these values for d^2 and h in Equation 8, and doing a little algebraic manipulation, we now have:

$(2\pi r/T)^2 = r\ a$ and solving for a,

$$a = r\ (2\pi/T)^2 \tag{11}$$

At this point Newton's followers had the formula for the acceleration of the moon toward Earth. It may seem odd that the moon accelerates toward Earth, and yet has a stable orbit. But acceleration can be non-zero, even when the speed is zero. When the clock pendulum swings, it has maximum speed and zero acceleration at the bottom, and zero speed and maximum acceleration at the top. When my car hits a brick wall, it has a huge deceleration (negative acceleration) and almost no speed. Newton had this concept all worked out.

Newton set the acceleration a = F/m where, he said, F = F_G is the gravitational force of the Earth-Moon system and m is the mass of the Moon, for it is Moon accelerating with its mass, M_{moon}.

Combining a = F/m and Equation 11,

$$a = r\left(\frac{2\pi}{T}\right)^2 = \frac{F}{m} = \frac{F_G}{M_{moon}} \tag{12}$$

The force, F_G is given by Newton's Universal Gravitation (Equation 4) and we have:

$$r\left(\frac{2\pi}{T}\right)^2 = G\frac{M_{moon}\,M_{earth}}{M_{moon}\,r^2} = G\frac{M_{earth}}{r^2} \tag{13}$$

Moving terms around, Newton's followers finally got Kepler's law of planetary motion as an equation.

$$\frac{r^3}{T^2} = \frac{G}{4\pi^2}M_{earth} \tag{14}$$

or in more universal terms we have,

The Newton-Kepler's Law:

$$\frac{r^3}{T^2_{orbit}} = \frac{G}{4\pi^2}M_{central\,body} \tag{15}$$

Compare Equation 15 with Kepler's results given in Table 1.

Somehow Newton knew he had the solution to Kepler's law, and stated such, but, as far as we know, he did not derive it algebraically, as we did here. But think of Newton's elation when he derived Kepler's law from his guess of Universal Gravitation. It is likely Newton fought his way through many dead ends, until finally he had the correct path to Kepler's universal law. It was a triumph of the mind that is difficult to overstate. Physics is always like this. Make a good guess; use mathematical rigor to see where it leads; check for consistency; make sure it corresponds with what previously passed experimental tests; devise new experimental tests; be as accurate as possible.

Now the full power of Kepler's law of planetary motion, and Newton's law of gravitation comes into play. We may use them to calculate the mass of Earth, Mars, Jupiter, Saturn, sun, or any body with a satellite having a measurable period of revolution and distance. Let's check it out.

Solving Equation 15 for the mass of the primary attracting body,

$$M_{central\ body} = \frac{4\pi^2\ r^3}{GT^2} \qquad (16)$$

The constant of gravitation, G, has been obtained by laboratory experiment. We need only measure the orbital period T, and orbital radius r, of a revolving body and we can calculate the mass, whether it is a planet, moon, or star. When the orbiting body is much smaller than that which it goes around, Equation 16 may be used directly. For a satellite, whose mass approaches that of the planet, the center of rotation lies somewhere between the two. Usually we can ignore that complication, unless high accuracy is required. Notice that the mass of the satellite is not in the equation and is therefore not needed. All satellites, no matter what their mass, have the same orbital period when the orbiting distance is the same. The distance to the orbiting body and the mass of the central body are all that determine the period of the orbiting body. That is the meaning of the constant value in column 7 of Table 1. That is also Galileo's discovery: all bodies fall with the same acceleration. It all comes together in Equation 15.

CHAPTER 6

Weighing Jupiter

We may use Equation 15, which is Newton's revision of Kepler's discovery, to calculate the mass of any central body, such as Earth, other planets, even the stars—provided they have at least one observable satellite, natural or artificial. We can also use Equation 15 to calculate orbiting distance, r, or orbital period, T. In the Earth-Moon system, Moon revolves about Earth with a period of 27.322 days (referenced to the stars) and orbits at an average distance of 237,280 miles. Taking care to be consistent in choosing the units of distance and time (from astronomical observation), we calculate and list planet masses and sun's mass in Table 2.

CENTRAL BODY	SATELLITE	SATELLITE PERIOD, T	SATELLITE DISTANCE, R	R^3/T^2	CENTRAL BODY MASS	CENTRAL BODY MASS
		days	miles		slugs	kilograms
Earth	Moon	27.32	237,280	1.79E13	40.5E22	5.94E24
Mars	Deimos	1.262	14,480	1.91E12	4.30E22	6.28E23
Jupiter	Europa	3.551	414,140	5.63E15	1.27E26	1.86E27
Saturn	Titan	15.945	763,660	1.75E15	3.96E25	5.78E26
Sun	Earth	365.26	92,346,000	5.90E18	1.33E29	1.95E30
Sun	Venus	224.7	66,790,000	5.90E18	1.33E29	1.95E30
Sun	Jupiter	4332	480,432,000	5.90E18	1.33E29	1.95E30

Table 2. Masses of four planets and sun calculated using Newton's revision of Kepler's law of planetary motion.

Despite the great variances of the planets' orbits, each gives the same value for the sun's mass as predicted by Newton and Kepler. This is a triumph of human reasoning. To be able, in effect, to weigh the planets and sun by observation of orbits and application of physical law is truly remarkable. No wonder the early 18th century was called the beginning of the "age of enlightenment."

We can arrange the terms of Equation 15 to see what it says about artificial satellites put into near-Earth orbit by high-speed rockets. When the Space Shuttle orbits 200 miles from Earth's surface, how long does it take to make each trip? We get from Equation 15:

$$T_{orbit} = \left(\frac{4\pi^2 \, r^3}{G\,M_{central\ body}} \right)^{1/2} \tag{17}$$

Adding 200 miles to Earth's radius of 3964 miles, r = 4164 miles = 6.700E6 meters. With Earth as central body (mass = 5.94E24 kilograms), Equation 17 gives the orbital period of 91 minutes for the Space Shuttle. On each trip the Shuttle goes a distance of $2\pi r$ = 26,163 miles in just 91 minutes, giving it a speed of 17,250 miles per hour or 27,755 kilometers per hour. This is why the Shuttle arches over into a more horizontal direction at high altitude. It must obtain a horizontal speed of just over 17,000 miles per hour to keep a constant r, and have a stable orbit.

Let us turn to Equation 15 again, and calculate what orbital distance a communication satellite must have to appear to stay stationary overhead, in what is called an Earth-synchronous orbit. We get:

$$r = \left(\frac{G M_E \, T^2_{com \; sat}}{4 \pi^2} \right)^{1/3} \tag{18}$$

For the communications satellite, T = 24 hours = 8.64E4 seconds. We can get Earth's mass M_E from Table 2. Solving Equation 18, the Comsat satellite must be placed 26,200 miles from the center of the Earth, or 22,236 miles above Earth's surface. The satellites of the Global Positioning System, GPS, are in half-synchronous orbit where they circle Earth every 12 hours. Using Equation 18, the GPS satellites should be 12,540 miles above the surface of Earth to work as planned. And so it goes; celestial mechanics is not much more than utilization of the Newton-Kepler laws of motion.

CHAPTER 7

What is Pulling?

It is now possible to find the analytical form for acceleration and close the loop on Galileo's measure of the acceleration of gravity at Earth's surface. We may combine Newton's laws of force and Universal Gravitation and find what's involved in the acceleration of falling. Newton taught that a force F will accelerate a mass m. Near the surface of Earth, the force that causes falling and acceleration is the gravitational force F_G. We are about to merge in an explicit way the force that causes acceleration and the gravity force to see what it is about Earth that gives the acceleration that Galileo discovered. We extend Newton's $F = ma$ (Equation 2) to also mean that the gravity force accelerates mass m in the same way. The source of the force does not matter; we can say $F = F_G$, where F is any universal force and F_G is specifically the force of gravity. This may seem like an unnecessary distinction, but it became a central point of Einstein's General Theory, as we shall see later.

Since $F = ma$ and $F_G = mg$, we may set mg equal to the force of Universal Gravitation, given in Equation 4. Putting these together;

$$mg = G(M_1 M_2/r^2) \qquad (19)$$

M_1 and M_2 are any two bodies separated by the distance r. So, at the Earth's surface, r is the Earth's radius, about 3940 miles; M_1 is, by definition, the mass m, and is our test body or Galileo's little cart; and M_2 must be Earth's mass M_E. Therefore the acceleration of falling, at Earth's surface is given by:

$$g = G(M_E/r^2) \qquad (20)$$

This equation can be generalized; M_E can be any attracting body: sun, Venus, Jupiter, a star, galaxy, or any chunk of matter. It will produce acceleration, g at its surface located r from its center. But it will be necessary to know the attracting body's mass to find the value of g. On the Earth's surface, r = 3940 miles, and we get Earth's mass from Table 2. Using Equation 20, these yield, g = 32 feet/sec-sec. This agrees well with Galileo's measurements.

Actually one has to be very careful to avoid circular reasoning. When G is calculated using Equation 5, the force must be independent of g. It can be difficult to design experiments that have truly independent parameters. Nevertheless, Equation 20 shows us how Galileo's acceleration depends on Earth's mass and size. If Earth were half its present size but had as much matter as it has now, g would be four times as large, and we would all weigh four times as much.

Where are we in our search for gravity and antigravity? Newton analytically described the force of gravity, and celestial mechanics is founded on scientific reasoning as well as careful measurements. The precision of space calculations, based on simple theory, is astounding. But we still come back to the basic question of "what is pulling?" Saying the word "mass" is not an answer, because mass is just a construct to help us give quantity to force and acceleration.

By 1700, science had reached its second plateau and got stuck for another 200 years. Physicists were busy studying all aspects of mechanics, heat, electricity, magnetism, and optics. Great strides were made, but in most cases, some form of action-at-a-distance was required to extend scientific discovery. To solve this riddle, physicists invented another word, "field," and the "theory of fields" had become the solution to "what is pulling?" It really doesn't provide a good solution to our fundamental understanding of gravity, but it does point in the right direction.

The idea of fields comes from the magnet. Just about every student has seen the demonstration of magnetic fields, where iron filings are sprinkled on a sheet of paper that covers a bar or horseshoe magnet. Tapping the paper lightly causes the bits of iron to bounce free of the friction of the paper and line up as tiny bar magnets under the influence of the "magnetic field." If done correctly, lines of iron filings connect one end of the magnet with the other. These lines are called "lines of force" that "reveal" the presence of the magnetic field. The lines go from one end of the magnet to the other end, forming a closed loop. Aluminum filings don't work. Wood sawdust doesn't reveal magnetic fields. But aluminum filings, sawdust, and everything else do get pulled toward the center of Earth. Every substance reveals the presence of the gravity field. Its lines of force are fairly straight, and each line of force originates at the center of mass of each body, then extends into space.

The theory of fields is a powerful tool and remains the cornerstone of teaching advanced physics. Unfortunately, it has become the litmus test for concepts in physics and blocks new roads to discovery. Even Einstein got mired in its seductive grasp, as we shall see.

CHAPTER 8

Gravity, Work and Energy

The most striking thing about the Space Shuttle is the spectacular blast of burning fuel during the launch phase. It takes energy, and a lot of it, to put a thirty-ton payload into orbit. Why? For the same reason it takes work or energy to push a cart up the hill or climb the stairs. "Work" is the name we give to the mechanical form of energy. Work is defined as force, F, times distance, s, (Energy = work = F times s), and has the unit of joule in the metric system, (pound-foot units in the English/American system). The force of one newton exerted over a distance of one meter produces one joule of energy. We can use this definition to express energy of a body in terms of its mass and speed because Galileo tells us how acceleration and speed are related. Since $F = ma$ (from Newton) and $s = \frac{1}{2} at^2$ (from Galileo), we put these ideas together to get:

Energy of Motion:

$$E = F\,s + mas = ma\,(at^2/2) =$$
$$\tfrac{1}{2}\,ma^2\,t^2 = \tfrac{1}{2}\,mv^2 \qquad (21)$$

This equation says there is energy of motion, called *kinetic energy*, when a mass is moving at speed v. When the direction of motion is stipulated, speed combined with direction is called *velocity*.

A small, half-kilogram satellite moving in low earth orbit at 27,775 kilometers per hour has a kinetic energy of $E = (0.5)(0.5$ kilograms$)(7715$ meters per second$)^2 = 1.49E7$ joules, or about 15 million joules. This is enough energy to heat 10 gallons of water from freezing to boiling.

There is an interesting relationship between momentum, mv, and motion energy, $\frac{1}{2} mv^2$. Whereas momentum tells us that a body moves in a specific direction, energy does not specify direction. A bowling ball has the momentum of its mass times its speed. The pins are knocked backward because the speed has direction. The energy of a body can always be referenced to any direction, at least in principle. It is possible, if not practical, to have the bowling ball ride the gutter, hit a sticky body, and have all the energy of the impact collected and focused onto the pins. Do it just right and all the pins go down. Score a perfect game with all gutter balls. The trick is how to collect and transfer energy. Energy is of fundamental importance in the understanding of our Universe.

Exactly what is energy? When I hold a barbell in my hand, I need a tight grip because of Earth's pull on the barbell. As I lift the barbell, my muscles do work, giving the barbell an energy equal to its weight times the height of the lift. My muscles provide this energy increase for the barbell, and because my muscles are not efficient, some energy is expended in my muscles, heating them up. I drive my car up a hill and increase the energy of the car because of the height of the hill. My car is not perfectly efficient, and the car's engine and drive train heat up. In both cases a mass, barbell or car, was given an increase in gravity energy by inefficient means. I drop the barbell, and it hits the floor. What happened to its gravity energy? The car rolls down the hill and is stopped with the brakes. What happened to its gravity energy? In both cases you need a thermometer to find out. The barbell and floor heat up a bit. The car's brakes get hot. The gravity energy went to heat. At some earlier time this energy was in the form of fuel, food, or gasoline. Earlier yet, it was solar heat-energy converted to plants and fossil fuel. What is heat?

By the middle of the 19th century, it was well established by careful measurements that energy is always traceable. It just moves from one form to another. It does not vanish. For example a cart (weighing 740 newtons) is pushed up a street 900 meters long on a hill that happens to be 107 meters high. If the cart pushes back with a force of say, 90 newtons all the way along the hill of 900 meters the work done is 90x900 = 81,000 joules. The energy it took from human muscle, horse muscle, or combustion engine has been converted to heat in the bearings and wheels, and most important to us, into the energy of its position. It is higher than it was, and has a stored (potential) energy relative to positions down the hill. When it is released, it will fall as Galileo's carts did, pick up speed (accelerate), and gain motion (kinetic) energy during the descent. If there were no friction, the kinetic energy at the bottom of the hill would be 79,180 joules. The missing 1820 joules are to be found in wheel and axle frictional heat.

When the cart hits a wall, all that energy will be converted to heat, warming the cart and wall. If the cart has brakes and the speed of the cart is held steady, the energy of falling will be transferred to the brakes in the form of heat. If the cart is made of wood, and it burns completely, it will release another form of energy exactly equal to the solar energy it took to form the wood during tree growth. Energy, in its various forms mechanical, chemical and heat, was fairly well understood by the 1870's. The science of thermodynamics explained the principles of conversion from one form to another. When all the atoms making up the cart have the common motion of going down the hill, we describe this motion energy as kinetic energy. The energy transferred to the brake shoe by friction is disbursed throughout the atoms of the brake, causing them to vibrate faster, and vibrate in random directions. Temperature is a measure of molecular vibration and is the relative measure of heat energy for any substance.

When we say that heat energy is in the form of vibrating atoms we are using classical mechanics as applied to atomic theory as if there were little springs connecting the atoms. This is all figurative speech. Even when we appeal to wave mechanics and talk about states of energy belonging to any ensemble of atoms we are still being figurative, although wave mechanics and quantum theory are powerful tools of understanding. The energy of atomic vibration is given the more formal name of *phonon*. All bodies above absolute zero temperature (-273.3C) emit and absorb photons. A body at higher temperature than its surroundings will emit more photons than it absorbs and in that way transfer its excess energy to surrounding bodies. Photons are energy. We can say that bodies store their received photon energy in the form of phonons. It is as if phonons are trapped photons.

The beauty of energy is its precise conservation. It never vanishes. Energy never appears from nowhere. It can always be traced. The equation of energy always balances, whether it is on the atomic scale, the human scale, or the cosmic scale. Conservation of energy is a powerful tool. It is the key concept for understanding gravity. The energy of gravitation is another form of energy. When the cart is at the top of the hill, it has energy greater than when it is at the bottom. All isolated systems of particles and bodies have a tendency to eventually reach a state of uniform energy. Energy transfers will occur until the energy is uniformly distributed. Unless its wheels are blocked, the cart will roll down the hill. If a hole opens to the center of Earth, the cart will fall into the hole, and eventually wind up at Earth's center. To understand gravity, we must explore the energy of gravity.

*   *   *

The Hidden Hand

When a ball is thrown straight up and the wind is not blowing, the ball comes straight back down. The path is vertical. Why doesn't it wiggle around, and maybe fall in a slanting path. If thrown straight up, it always comes straight down. Sure, it is reacting to the gravity force, but why is there such an economy of motion on every throw? Energy would be conserved independent of the path, but motion always follows the path of least resistance. Is there a way to describe this "economy?" Funny you should ask.

William Hamilton, a brilliant Irish mathematician, building on the work of P. Maupertuis, an 18th century French mathematician, developed this economy concept we now call *least action*. Hamilton, at age 28 in 1833 realized that the study of light would reveal how matter particles move from place to place:

> *The answer, I think, must be the principle or law, called usually the Law Of Least Action; suggested by questionable views, but established on the widest induction, and embracing every known combination of media, and every straight, or bent, or curved line, ordinary or extraordinary, along which light (whatever that may be) extends its influence successively in space and time…*

By mid 19th century, Least Action was expressed as the product of a body's momentum and displacement, or the product of its energy and time of travel. This is an empirical law, as is most of physics. It comes from observation. Every measurement, taking into account all the constraints on motion, always shows that motion has the least amount of action. For any change in energy of a body, its motion through space occurs in the least amount of time. It is as if there is a hidden hand guiding the motion to achieve highest efficiency. As a body falls to Earth, Least Action guides its motion at all times, even over minute portions of its path. If the ball is thrown up and hits a ceiling, the ceiling constrains the motion, but Least Action is always at work. The path is always the most efficient.

If each point in space is active in terms of action, then we may assign a property to space that determines the motion. And by the late 19th century this property got the name of "field" such as gravitational field, or electric field, or magnetic field. But remember the word " field" is a mathematical way of describing the hidden hand. What is it about space that regulates all motion—all changes of state, such that action has optimum economy? Matter responds to force. Does space exert force to guide matter? How can space supply the force necessary to achieve least action? We explore these ideas in Part II. But first we need Einstein to explain mass, and we must explore the structure of matter.

CHAPTER 9

Mass, Energy, and Dr. Einstein

In 1905, Albert Einstein, while working in the patent office in Berne, Switzerland, published his now famous paper on "Special Relativity" that he called *On the Electrodynamics of Moving Bodies.* By assuming that the speed of light, c, in empty space always has the same value however it is measured, he derived that when a chunk of matter has its energy increased relative to another chunk of matter, it appears to have more mass. The added energy can be kinetic or thermal. He concluded that mass and energy are closely related and the relationship is a function of the speed of light.

The speed of light has two interesting properties; it has great magnitude, and it is independent of the frame of reference in which it is measured. Many kinds of measurements of light speed have been made, with the presently accepted value of 2.9979 ± 0.0001 E8 meters per second (186,000 miles/second).

Einstein, at 26 years old, showed that since the speed of light is constant, the dimensions of space and time are relative to the particular frame of reference in which the measurements are made. When I am in my car, traveling 60 miles per hour, all objects outside the car, are less in size in the direction I am going. But this speed is so slow, compared with the speed of light (186,000 miles per second), that everything looks the same as when I am stopped. When I am in my rocket ship going 52,000 miles per second, objects outside my ship, appear to have shortened by 4 percent in the direction I am going. And just as Einstein said, my clock is running 4 percent slower than the clocks outside. Space and time are not absolute. They depend on relative speed. I accelerate my rocket ship to a speed of 185,000 miles per second. Now all outside dimensions are only 10% of what they were, and my clock has slowed in like manner.

Let's look at Einstein's time dilation formula, and see how it impacts Kepler's equation of orbital motion. You and I get in my rocket ship and blast off in the direction of Moon's orbital axis toward the star Delta in the constellation Draco. Burning lots of fuel, we accelerate to a speed of 161,000 miles per second. Your job is to measure the Moon's distance from Earth and its period of revolution using our onboard clock. My job is to avoid space debris and worry a lot.

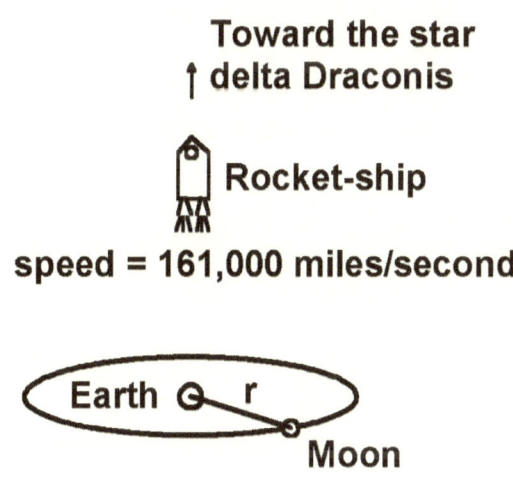

Figure 6. We speed away from Earth and Moon at 161,000 miles per second, and as we go, we measure Moon's orbital period, and its distance from Earth. We verify Einstein's theory of relativity, and learn that mass is energy.

A few weeks into our trip, I check with you to see how things are going. You report that the Earth-Moon distance is the same as before, 237,280 miles, but the Moon is taking twice as long to go around Earth as it does when we are sitting on Earth. So we pull out the time dilation calculator to check it out. Einstein in 1905 said:

$$\textit{Time Dilation}: \quad t_R^2 = \frac{t_{E/M}^2}{\left(1 - \dfrac{v^2}{c^2}\right)} \qquad (22)$$

Our rocket ship clocks measure t_R and Earth clocks measure $t_{E/M}$. We checked all clocks just before launch. Clocks on the rocket ship are working just fine, as are the clocks on Earth (we assume). Since $v/c = 161/186$, you calculate about twice the Moon's orbital period, T_M, (the Moon's proper period of 27.36 days). You get 54.7 days, according to Einstein's time equation. Your calculation and your observations match up just fine. And because we are traveling perpendicular to the Moon's orbit its dimension is the same as when our speed was zero relative to the Earth/Moon frame of reference. So far, so good.

Now if we apply Kepler's law, what will it say about this phenomenon of relativity. Let's follow though with the algebra and see what develops.

From Kepler/Newton, we write Equation 15 (see "Moon is Falling"), but now from our reference on the rocket:

$$\frac{r^3}{T_R^2} = \frac{G}{4\pi^2} M_E \qquad\qquad (15R)$$

Time $t_{E/M}$ = the Moon's period T_M as measured by Earthlings. Time t_R becomes Moon's period, T_R as measured by you in the speeding rocket. Substituting for T_R in Equation 21 by using Equation 15, we have:

$$\frac{r^3}{T_M^2} = \frac{G}{4\pi^2} \frac{M_E}{\left(1 - \dfrac{v^2}{c^2}\right)} \qquad\qquad (23)$$

The last term of this equation may be considered the Earth's effective mass, as observed from our high-speed rocket. The time dilation term appears to dilate Earth's mass. Since the Moon appears to slow down to give twice the period—and the orbit's radius has not changed—we must conclude that the Earth's mass has doubled. Earthlings of course feel no such effect. It is an illusion of relativity. But we press on. Let's call the last term $M_{E,R}$, the Earth's mass as seen from the rocket, and do some algebraic juggling and see what happens. Remember, mass times the square of speed is energy.

$$M_{E,R} = \frac{M_E}{\left(1 - \frac{v^2}{c^2}\right)} = \frac{M_E c^2}{c^2 - v^2}$$

$$M_{E,R} c^2 - M_E c^2 = M_{E,R} v^2 \qquad (24)$$

$$Energy = M_{E,R} v^2 = \left(M_{E,R} - M_E\right)c^2 \quad or$$

$$Energy = \Delta M c^2$$

The mass of Earth as observed from the rocket minus the Earth's mass as measured by Earthlings is an apparent increase in mass. We use the symbol, Δ, with M to indicate this change in mass
.

This set of equations shows how time dilation gives the equivalence between energy and change of mass, ΔM (= $M_{E,R}$ - M_E). This is Einstein's most famous equation; *energy is equal to mass times the speed of light squared*, $E = mc^2$. The important point here is that energy is disguised as mass. We tend to think of mass in the Newtonian sense, as if the tiny sub-atomic particles are some kind of hard little spheres. There are no hard little spheres. All matter is energy. When we observe our world and all its variety, we are looking at variety in the spatial configuration of energy. Substance is energy. My hand is a complex spatial configuration of energy. A rock is another complex spatial configuration of energy. All elements of matter are configurations of energy and

Einstein's formula tells us how to convert from the dimension of mass to the dimension of energy.

How significant is Einstein's mass-energy equivalence? If I hit a baseball and transfer the energy of my muscles to the bat, and then to the ball, the ball flies away with a tiny bit more mass from my reference point than it had when lying still in the Umpire's pocket. If I add heat to the ball, it will have a tiny bit more mass relative to its cooler surroundings. The extra energy, kinetic or heat, manifests itself as increased mass in the Newtonian sense of mass. How big is this effect? It all depends on which side of the ball you are on, so to speak. It takes a huge amount of energy to measurably increase the mass of a ball. And conversely, a very small amount of mass represents a huge amount of energy. Einstein's famous equation gives us the relationship between mass and energy.

Let's try it out. I hit the 500- gram (about 20 ounce) ball, and knock it into low Earth orbit with a speed of 17,250 miles per hour (27,775 kilometers per hour). From $E = \frac{1}{2} mv^2$, we find it has an energy of 1.49E7 joules. This increase in energy is accompanied by an apparent increase in mass, given by $m = E/c^2 = 1.49E7 / (2.9979E8)^2 = 1.66$ E-10 kilograms or approximately 0.16 micrograms. The mass of the ball would increase by only .000000032 percent.

Now lets get on the other side of the ball. If I could convert but one millionth of the ball's mass (500 micrograms) into energy, how much water could I raise in temperature from 0 C to 100 C. I get $E = mc^2 = (500$ micrograms of ball) times $(2.998E8$ meters/sec, speed of light$)^2 = 4.5E10$ joules which equals 1.08E10 calories. Each calorie can heat one gram of water 1 degree C. So the 1.08E10 calories will increase the temperature of 108 thousand kilograms of water by 100 C. That is a whopping 25,000 gallons of water, brought to boiling, by a microscopic bit of ball converted from mass-energy to heat energy. What is going on here? What is it about matter that it has so much energy packed into it?

We must inquire into the structure of matter to understand where the mass-energy lies, because gravity depends on the way in which mass-energy is distributed. We are after the fundamental particles of matter and their distributions of energy because energy distribution is gravity. We will examine the three constituents of ordinary matter: photons, protons, and electrons.

CHAPTER 10

Unpacking the Atom

Our most familiar form of energy is light. It is energy that is never at rest and has no mass that you can put on a scale. Yet light particles, called photons, have an energy-equivalent mass according to Einstein's relativity. Photons interact strongly with matter, exhibiting reflection, absorption, refraction, scattering, and other effects. In a sense, photons get trapped by matter, raising its energy content. Put the pot on to boil, and the heat source (electric stove coil or burning gas) radiates a lot of photons to the pot. The photon energy gets absorbed, and the temperature goes up. Where do the photons reside?

The classical view is this: The metal of the pot consists of atoms of iron, copper, aluminum, or perhaps some other metal. Each atom of the metal has a very small nucleus of protons and neutrons, surrounded by electrons that together structure the atom's local space in a special way. The formal name for the study of this structure is "wave mechanics," because physicists use equations that describe wave motion to represent matter or energy. These matter or energy waves are standing or stationary waves. When I pluck a guitar string, the vibrating string is pinned at both ends holding the waving string stationary at its ends. Its energy of vibration is transferred to wood, air, and ear, and the string soon stops vibrating as it loses its energy. This wave nature of matter-energy was invented in the early twentieth century and is part of physics called *quantum mechanics*. The atoms, singularly, in pairs, or in groups are said to occupy well defined states of energy and jump from one state to another absorbing or emitting discrete bits of energy, usually in the form of photons.

Back to our pot. The photons add to the electron energy waves of the metal. These electron waves are coupled to the atom's nucleus by a force called the electric force. Unlike pure attraction which gravity seems to exhibit, the electric force is either attractive or repulsive. Physicists have devised a way to describe this property of attraction and repulsion. We say that the electrons and protons have charge, and it is charge that causes attraction or repulsion. These are just a bunch of words with which we pretend to know what is going on in the atom. Actually, it is amazing how much can be done with these primitive notions of atomic structure. Most of our advanced electronics technology has been developed on simple ideas of charge and wave mechanics. When light energy is absorbed, the photon energy waves harmonize with the electron waves, and new waves of higher energy are formed in the atomic structure. The charge force couples the electron wave energy to the nucleus, and it vibrates relative to the matrix of atoms composing the metal. The energy is distributed to electrons and nuclei, and the entire structure is in a higher energy state. As random events, photons will be reformed, go off at the speed of light, and be reabsorbed somewhere else in the chunk of metal pot. Photons are continuously emitted and absorbed. When the water in the pot boils, steam takes away the energy as fast as it comes in from the stove. The system of stove, pot, water and steam is in equilibrium. Photon energy in equals heat out. Shut off the stove, and photons will continue to be emitted by the pot, lowering the energy state of the metal, until equilibrium is again established at room temperature. At equilibrium (when the pot cools to room temperature) with its surroundings, photons going in equal photons going out.

Photons are bundles of energy that travel through space at 3 E9 meters per second, which is the speed of light. Photons that carry the information of a television signal have energies in the E-25 joule range and photons that create a sensation of light in the eye-brain system have energies near 4E-19 joules. Gamma ray photons have energies ranging from E-13 joules to E-9 joules. Photons are the free energy of the Universe. They are pure energy in motion.

This process of energy transfer by photons is familiar to everyone. Our bodies are energy machines converting sugars to heat by a well-controlled oxidation process in every body cell. This process is designed to keep the vital organs close to 98.6 degrees Fahrenheit or 37 degrees Celsius. It takes on average, 2000 food calories each day to maintain this healthy temperature. Each food calorie equals 1000 heat calories. It therefore takes 2 million heat calories a day to stay healthy. This heat is equal to 8.3 million joules of energy. Since there are 86,400 seconds in a day, the rate of energy usage is 8.3 million joules divided by 86,400 seconds, which is 96 joules per second. One joule per second is defined to be one watt of power. The average human body is a heat machine running at the rate of 96 watts. Each of us is generating and emitting nearly the power of a 100 watt light bulb all the time we are alive. No wonder a crowd of people can heat up a room.

Sit by yourself in a cold room wearing summer clothes. Your skin is at approximately 92 degrees Fahrenheit when you enter and radiates photons of energy to the cold walls. If the walls are at only 42 F, they are returning only half the energy your body is radiating. Most of the photons emitted from the surface of our skin have energy of 2E-20 joules. Your skin cools, and your body makes adjustments to maintain its core temperature, lest you die. The blood circulation to your limbs is diminished, and you shiver. The oxidation process in your cells can not increase much, otherwise there will be cell damage. Your heat engine is in danger. You put on thick clothing, and the heat radiating as photons from your skin raises the temperature of the air trapped in the cloth. The warm air layer surrounding your body radiates photons back to your skin. A new equilibrium is established, and your body can return to normal function.

Photons are energy. When they are absorbed in the energy wave structure of atoms, the matter has increased energy. Energy is mass. The photon-absorbing body of matter will behave as if it has increased mass. What can we say about the electrons, protons, and neutrons themselves? What are they? Are they energy too? When a neutron is knocked out of a nucleus in some high-energy event, as in a particle accelerator, the freed neutron decays into a proton and an electron. The mean lifetime of a free neutron is 14 minutes, meaning that half of any collection of free neutrons decays in 14 minutes. The only stable fundamental particles comprising all known matter are the electron and the proton. The photon is also stable, but has the special property that it is always traveling at the speed of light. Photons, electrons, and protons have strong interactions, suggesting that their energy-wave properties are harmonic resonators. It is as if the atom is a collection of tiny strings of oscillating energy, all in some kind of harmonic resonance. Their energy structures must link in many specific phase relationships. Fortunately it is not necessary to work out all this complexity to solve the problem of gravity. It is the basic energy structure of a proton and electron that gives us gravity, but in a surprising way.

By the end of the 19th century, with the discovery of radioactive decay as a new source of energy, the conservation-of-energy principle seemed to be violated. Particles were being emitted with high kinetic energy from atoms of radium, uranium, and other elements. The source of the energy was unknown. Was this a violation of energy conservation? Physicists discovered that the high-energy emissions were photons, protons, neutrons, electrons, and even helium nuclei of immense energy. Further studies by high-energy physicists revealed that extremely high-energy protons from deep space traveling at very nearly the speed of light were hitting Earth's atmosphere and creating showers of particles and gamma rays. The sequence of decay for these "cosmic rays" showed transmutations from proton to gamma rays, back to protons, including an entire zoo of new exotic particles. As the measurements improved, the principle of the conservation of energy was found valid for all events.

Protons were collided head-on with anti-protons in high-energy collider machines, creating gamma rays. Experiment after experiment demonstrated that protons and electrons can transform to photons, and photons with sufficient energy will transmute to pairs of protons or pairs of electrons. Experiment upon experiment showed that the distinction between mass and energy had become a blur. Einstein's mass-energy principle was verified time and again at the sub-atomic level. Physicists measured the mass equivalent energy of the sub-atomic particles. The electron at rest has a mass equivalent energy of 8.187E-11 joules, and the proton at rest has a mass equivalent energy of 1.5033E-10 joules. A more common term for particle energy is the electron volt written as eV. One electron volt equals 1.602E-19 joules. Thus, the rest energy of an electron is 0.511 million eV, and that of a proton is 938.3 million eV. Mass is energy indeed, but what is the source of this energy?

When a proton or body of protons is moving at a speed v, relative to a frame of reference, it has kinetic (motion) energy in addition to its mass-energy, mc^2. In modern relativistic terminology the total energy, E is expressed as a sum of motion energy (momentum p times the speed of light c) and the Einstein mass-energy.

$$E^2 = p^2 c^2 + m^2 c^4 \qquad (25)$$

This expression applies to any kind of particle or collection of particles. Using relativity this way, the mass-energy of a particle is defined with the mass at rest, relative to the measurement apparatus. The mass-energy is precisely rest energy. The momentum term, p^2c^2 takes into account kinetic energy because $p = mv$. For the proton, mc^2 is 938 MeV, whatever is its speed. Some authors call E the *momenergy*.

A particle, modern relativists say, does not gain mass with speed; it gains energy. The distinction between mass and energy has become a blur. Mass is a term of convenience. When talking about subatomic particles such as electrons, protons, and photons, the concept of mass as having geometrical structure leads only to confusion. This sad state of affairs is a result of treating reality as consisting only of particles. When physics ignores the role space plays in particle mechanics, mass and energy are difficult to define, and the relationship between the gravity force and other forces can not be obtained.

Proton and electron are photon-like energy in disguise. All of matter is a complex configuration of energy. Fortunately, we do not need to figure out the complexity of matter to understand the basics of gravity. Einstein based his theory of gravitation (General Relativity) on matter being able to change the geometry of space. We can ignore the details of his space-tensor theory, and still extract the final clue we need to find the secret of gravitation.

CHAPTER 11

Elevators in Space

It occurred to Albert Einstein, while teaching in Zurich in 1912, that a ride in an elevator shows the equivalence of gravity and inertia.

> We enter what looks like an ordinary elevator. While the elevator is sitting still, we feel a force on our bodies: our weight, due to the pull of gravity. As long as we feel our full weight, we think we are stopped somewhere in the elevator shaft. A sleeping gas enters the elevator. As we sleep, the elevator is moved far from Earth and is accelerated by a rocket to a value equal to g, the acceleration of gravity at the Earth's surface. We wake up and ask the operator what floor we are on. Without windows, we can not see that we are far out in space. Acceleration gives us the illusion of gravity.

This equivalence concept led Einstein to develop the General Theory of Relativity he published in 1916 while doing research in Berlin. Space, he said, has a property that transforms inertia to force. Matter, because of its mass-energy, bends space giving it curvature. We are not aware of space curvature, since we live where space is weakly curved and our local space seems to be a 3 - dimensional space described by Euclidean geometry. We do not experience the curvature except by gravity acting as a force field.

The result of his very complex multi-dimensional theory is that local curvature of space, symbolized by the Greek letter sigma, σ, is proportional to the local mass-energy. He wrote this in a differential form, where a small change in curvature, $d\sigma$ is given in terms of a small change in energy, dE:

Change in space curvature = $d\sigma$ = K dE   (26)

As we approach a mass such as Earth, we enter space of increasing curvature. A mass as large as Earth will significantly curve space, and there will be a radial gradient of curvature. This means the curvature will increase the same from any direction, toward Earth. An analogy is to represent space as a rubber sheet supported horizontally, like a drum membrane, in an Earth laboratory. An object placed on the rubber sheet will depress the rubber beyond the object itself. A small ball, placed on the rubber sheet, will roll toward the object because of the depression of the surface. Einstein does not explicitly tell us how Earth, or any other mass, curves space. It is hidden in the geometry he used. Einstein's formulation is overly complicated—for what we need—to find the source of gravity. And there is another problem with his curvature equation. K is an unresolved constant. He shows how K is related to Newton's G, but both K and G are not just numbers. They are also physical quantities that must depend on something more fundamental, which is what we must find. Luckily, it just takes turning Einstein's theory around a bit to find out what is really curved.

If we divide both sides of Einstein's equation of curvature by a small increment of radial distance, dr, we will have:

Differential Curvature $= d\sigma/dr = K\,(dE/dr)$ (27)

dE/dr is an energy gradient (E changes with distance r). Since energy is force times distance we can say $dE = F\,dr$, and dE divided by dr (written as dE/dr) is a force. Rearranging the terms of this equation:

Force exerted by curved space =

$$F = dE/dr = (1/K)\,(d\sigma/dr)   (28)$$

Now we have force, expressed as the ratio dE/dr, equal to a constant, 1/K, times the change in space curvature with radial distance. So, for Einstein, gravity was a response to the changing curvature of space, and K is a measure of the stiffness of space. In Einstein's space, bodies always move on *straight lines* that are tangential to the curvature of space. In our observable three - dimensional space, we see curved motion such as the orbits of satellites and planets, as well as falling apples.

Einstein called these observed trajectories *illusions,* caused by the curvature of space. Is space just geometry? Is the force pulling me into my chair an illusion? I think not. The key to our search is in Equation 28. Gravitational force is the radial change in energy given by dE/dr. Einstein chose to pursue a space geometry where a curvature is the result of the mass-energy confined in matter itself. But the gravity force is given by dE/dr, which means energy of some sort is changing with distance.

We say this is an *energy gradient,* by which we mean, that energy decreases moving away from a central body of matter. What energy is decreasing? Now we have come, finally, to the central crisis of gravity. Gravity is a property of energy, but of what energy? Twentieth century physics avoided this question and deals only in particles and their fields of force. Mass has its force field and we call it *gravity*; charge has its force field and we call it the *electromagnetic field*; protons and neutrons are influenced by a very short range force field we call the *nuclear force*. Space is populated by fields and each field acts on its own kind of particle. What a mess. Then to top it off, standard physics searches for ways to unify these fields, as if they are all really the same basic thing in different disguises. Confusing as all this field structure is, it works pretty well. Field physics becomes field engineering and useful products such as electric motors, television, and cell phones are designed and made. All this concrete usefulness convinces us there is some basic underlying phenomenon that explains the workings of the Universe. The key to understanding the structure of our Universe is to structure our physics from the point of view of energy variations. It will be a lot easier if we think of particles as *energy condensations*, and fields as *energy variations*.

The multi-dimensional formulation of Einstein's General Relativity is essential however, when it is necessary to consider large portions of space, speeds approaching the speed of light, and the extremely high densities of mass-energy that exist in neutron stars and black holes. The one other result of his theory of General Relativity we do need for this simple approach is that the *Universe is finite but unbounded*.

We live in a three-dimensional space: left-right, forward-backward, and up-down. So naturally, we expect that if we travel in a straight line in space, we will eventually come to the end of a finite Universe, whatever that may look like. But that *end* would be a boundary, and Einstein taught that the Universe has no boundary. We can not come to an *end*. Traveling long enough, we would return from whence we started traveling that supposedly straight line. By Einstein's theory, lines are straight only in relatively small portions of space, where the mass-energy densities are not too large. His theory teaches that on a large scale, space is curved; and all inertial motion must follow the lines of space. Earth's surface is somewhat of a bounded but endless space. In an earlier century when only surface travel was possible, one could have traveled round and round the world almost without impediment. The end of travel would not have been in sight, but the traveler was constrained to Earth's surface. If the surface were a perfect sphere with no topography and the travelers were completely unaware of up or down, the surface becomes ideally open to free travel and the traveler is not aware of being bounded. Such is our Universe. We feel free to travel in three dimensions unaware of any cosmic boundaries.

We can not possibly know what space looks like as a whole, because the fastest transfer of information we know of is light, and it travels at only 186,000 miles per second. The photons from very distant galaxies have been on their way for over 10 billion years. We are looking into the past when we observe the starry sky. It's the best we can do. But it is good enough.

Thanks to Einstein showing that we live in a finite Universe and thanks to astronomers who have taken the measure of space, we can find the source of gravity. The fact that our Universe is unbounded and has a finite size will enable us to evaluate Newton's constant of proportionality G and scale gravitation absolutely. This is our next task.

Part II

FINDING THE SOURCE

What we look for is sometimes too big to see.

We must venture beyond the known ways of understanding our Universe if we are to find the actual cause of gravity. In doing so, however, we climb onto the shoulders of those giants of physical science, Galileo, Kepler, Newton, and Einstein. From this lofty perch we find our direction and move beyond today's plateau of knowing. Einstein has given us the final clues we need to crack the code and solve the riddle of gravity. The answer lies in the unique relationship between the mass-energy of a sub-atomic particle and the structure of the immense Universe.

The key to understanding gravity is to grasp the intimate connection between the incredibly small particles that make up the Universe and the energy distribution of the entirety of observable space. The Universe consists of space and energy, nothing more. When we say space we really mean *space-time* since we experience changes as events in the distribution of energy throughout space. We define *time* as ordered changes of events. Energy is distributed throughout space as particles (protons, electrons and photons) and as a *continuum of energy* that we can call *space-energy*. The protons and electrons are sub-atomic concentrations of energy, a discrete amount of energy packed into a tiny volume of space. The proton and electron are minute dots of concentrated energy in a vast sea of space-energy. Photons are also concentrations of energy, but are unique in that they travel at the speed of light. The Universe consists primarily of space-energy, proton energy, electron energy, and photon energy. The proton, electron, and photon are condensations of space energy. A small portion of space energy is condensed in each particle. The seemingly infinite variety of matter, found in the Universe, is constructed from these four forms of energy.

Many other particles have been observed, but with few exceptions only the electron and proton, which make up ordinary matter, are stable. Physicists do observe transformations from photons to electrons plus anti-electrons, and from photons to protons and anti-protons. And, in collisions of electrons with anti-electrons, and protons with anti-protons, high-energy photons are formed. These transformations do not involve any change in space-energy. If all protons and electrons were derived only from conversions from photons, we would expect to see an equal number of anti-electrons and anti-protons somewhere in the Universe. But very little anti-matter has been observed. This lack of anti-matter has been a difficult problem for physicists because they base particle theory on local conversions of free energy and neglect the role of space-energy. Our concern is with particles formed from space-energy. The Universe also abounds with neutrinos, small bits of mass-energy that surge forth from the nuclear furnaces located inside of stars. They react with almost nothing. The role of neutrinos is yet to be adequately understood.

CHAPTER 12

Charting the Unknown

The proton and electron are the building blocks of matter. Each substance is a complex configuration of electrons and protons representing legions of different energy states. It is not necessary to delve into this complexity or to further split the particles to understand gravity. Quarks and force-exchange particles are the tools of the particle physicist and can only obscure our path to gravity, which is very simple.

Space-energy is nearly uniformly distributed in space. This uniformity cloaks its existence, because only *variations* in space-energy play an active or observable role with protons, electrons, and photons. It is sort of like a skater on a frozen lake. It doesn't matter whether the ice is 3 inches thick or 300 feet thick, as long as the surface is smooth and level. Skating is only possible if the skate will slightly depress the surface. The little depression beneath the well-sharpened skate gives the skater control. It is the small variations in the smoothness of the ice that are important and measurable. So it is with space-energy. Only its variations result in real forces. The amount of space energy is not important to the structure of the Universe, but its variations are everything.

Variation in space-energy is the source of gravity.

It is variation of space-energy that manifests itself as force, the gravity force. And those thinkers who have scaled the Universe have given us the necessary concepts to derive its basic structure and to take its measure.

Thanks to Galileo, Kepler, Newton, Einstein, and many others, we now have the tools we need to break the gravity code. What ideas have they put in our tool kit? Briefly:

- Motion always exhibits least action, no matter what the constraints.
- Energy is always conserved.
- Mass is energy.
- Force is a gradient of energy.
- The gravity force decreases with distance r, as $1/r^2$ and the gravity potential energy decreases as $1/r$.
- The energy density of the Universe is uniformly distributed.
- The Universe is finite, but unbounded.

CHAPTER 13

Our Compass

Using these ideas as our guide, we develop a model for the formation of a particle from space-energy and use mathematical logic to see where it leads us.

> Each particle is a condensation of space-energy. The amount of space-energy condensed to a particle is exactly equal to the particle mass-energy, which physicists have measured in the laboratory. The gravity potential, discovered by Newton tells us what the energy profile is. We mathematically add all the space-energy used for any particle everywhere in the observable Universe and equate it to the particle's mass-energy. Since space is finite, the sum of energies is finite, giving us the exact energy profile at any point in space. This method of reasoning gives us what we seek: the true energy profile of the space-energy for each particle. The energy profile, which is the change of space-energy with distance, is the gravity force.

Albert Einstein, in a lecture at the University of Leyden in 1920, gave us the root of this idea.

> "Since according to our present conceptions the elementary particles of matter are also, in their essence, nothing else than

condensations of the electromagnetic field, our present view of the Universe presents two realities which are completely separated from each other conceptually, although connected causally, namely, gravitational ether and electromagnetic field, or - as they might also be called - space and matter."

The root idea in this statement is the *connection between the gravity force and a field*, except we replace *field* with *a change of space-energy with distance*. By describing field in terms of energy, we remove all the encumbrances of having to deal with field theory.

Now, Einstein's statement would read:

Since, elementary particles of matter are, in their essence, nothing else than condensations of *space energy* our present view of the Universe presents two realities which are completely separated from each other conceptually, although connected causally, or—as they might also be called—*space energy and matter energy.*

Astronomers have found that, taken on a large scale, the matter of the Universe is uniform. This means that the energy density used to form all matter is also uniform. This fact enables us to express the gravity force in terms of the size of the Universe and all the matter that is in it.

Protons and electrons are energy condensations, having been formed from the energy of space shortly after the beginning of the Universe. And since energy must be conserved, the energy packed in a proton or electron left a depletion of space energy, exactly equal to the energy of the particle. Energy is conserved. Each proton and electron have a complimentary depletion of space energy. As the Universe expanded, each depletion stretched out to extend across the dimensions of the observable Universe, a distance estimated by astronomers to be approximately 12 to 13 billion light years, or 1.2 E26 meters.

Gravity is a property of this space-energy depletion, and the way it works is quite simple. The space-energy that surrounds each bit or body of matter is not exactly uniform, and the measure of this non-uniformity is the gravitational force. Gravity is not mysterious action-at-a-distance. And it is not the bending of space-time geometry, caused in some mysterious way by matter. It is not a mystery. It is the variation of space-energy—demanded by conservation-of-energy—which sets up the conditions for gravity to exist. Gravity is simply the response of matter, which is condensed energy, to any space-energy non-uniformity. We need to generate the mathematical structure of the space-energy depletions to fully understand gravity, as well as find ways, if possible, to achieve gravity shielding and anti-gravity.

Try this mental picture to get a better idea of what is going on. Consider energy to be uniformly spread throughout the observable Universe, as if it were some kind of special gas. A particle of matter comes into being when a portion of this energy-gas is condensed into an extremely small volume. This particle of matter is a small amount of space-energy gas packed into a tiny bit of space-time. The remaining space-energy gas density far from the particle is disturbed very little. As we approach the particle from a distance the space- energy gas density decreases. This rate of decrease of space-energy can be written as the gradient of energy, which is the change in energy density per unit of distance. We can write this as the ratio $\Delta E/\Delta r$, which is, precisely, the force we call gravity. (The notation $\Delta E/\Delta r$ means a change in energy, ΔE, divided by a small change in distance, Δr. This can be read as delta energy divided by delta distance. Mathematicians use the delta notation to mean difference or differential.)

Or try this picture. Pretend that geometrical space is in two dimensions, like the flat surface of a level desert. The vertical dimension seems to be sand, but the sand grains are bits of energy extending for as far as the eye can see in all horizontal directions. The vertical dimension is energy represented by the sand. Locate a point on this surface. Now dig up the sand in the following way. Scrape off a little sand as a circular ditch when far from the point, and stack it at the point. The pile of sand-energy (the particle) is stable. As you move closer, dig deeper, such that you dig a depth that goes as the inverse distance—when you are half the distance you started from, you dig twice as deep. As you approach the marked point your pit is quite deep and the pile of sand/energy at the center point is very high. No sand-energy is lost. You have made a sand-energy pile, and a sand-energy hole associated with that one pile. Gravity is the steepness of the remaining sand-energy in the hole surrounding the pile. Far away, the slope is shallow and gravity is correspondingly weak. Near the marked point, the pit wall is very steep and gravity is much stronger. Where the pit meets the pile, there is a smooth transition of energy from the pit to the pile. The pile represents the particle's mass-energy, which is confined to a miniscule portion of space. The steepness of the energy pile is extreme and represents the very strong forces of the nucleus of atoms. But the pit spreads to the edge of space giving a very gradual change of energy and the weak force we call gravity.

Electrons and protons are stable. Other particles may be made by the condensation of space-energy, but they are not stable and disintegrate into smaller bits of matter-energy. We are interested in only the effects of stable condensations of energy since the Universe is almost entirely made of stable matter-energy. The slow evolution of stars is witness to the stability of protons and electrons. Since the electron and proton exhibit an extremely powerful force—the electrical force—we will confine ourselves to electrically neutral matter, where for any piece of matter there is exactly an equal number of electrons and protons. We will come back to this issue of electrical effects in Chapter 21.

The gradient in the space-energy, associated with each particle, is the source of gravity, because an energy gradient (change in energy with distance) is what we experience as force. The very formation of matter-energy from space-energy has given us our gravitational force. The formation of matter is the formation of gravity.

Your body is composed of equal numbers of electrons and protons, each of which has its own depletion of space-energy. These depletions combine, so that your body as a whole has a space depletion of energy. Earth is composed of equal numbers of protons and electrons, which taken collectively has a huge depletion of space-energy. Earth's depletion and your body's depletion compete for the available space-energy in the local Earth-person space, creating a mutual force of attraction we call gravity. Gravity is the interaction of depletions of space-energy associated with "rigid" bodies of matter. We call bodies rigid when the electric force holding their atoms together is very strong, as it is in rock and steel.

From these simple ideas, we can create an equation for the gravity force (between any collections of matter) that has no arbitrary constants, but instead shows a dependence on the size of the Universe as well as the amount of energy it took to form all the matter in the Universe. We can do this because the energy of all the protons is the same, as with the electrons, and the Universe is finite. Since the energy comes from no farther than the size of the Universe, we can determine the shape and magnitude of the space-energy-depletion, with no hidden variables. We can find gravity.

CHAPTER 14

The New Gravity Equation

We present here a new gravity equation for examination. This gravity equation is easy to derive, because it is a simple addition of energy. But before we get into the details of the derivation, let's take a look at this new result.

The New Gravity Force:

$$F = \frac{2N_1 N_2 E_n L}{3N_U r^2}$$
newtons (29)

F is the force of gravity between two bodies or collections of matter, one with N_1 particles and the other with N_2 particles, where each particle is a nucleon: an electron coupled to a proton. By standard definition in physics, a nucleon is a proton or a neutron, as found in the nucleus of an atom. For gravity considerations, the particles of interest are electrically neutral, and have a mass-energy approximately that of a proton plus an electron. We will use the term *nucleon* to mean the neutral particle, having a mass-energy, designated by E_n, nearly equal to that of a proton. The Einstein mass-energy of a nucleon is about 939 million electron volts of energy which is 1.5 E-10 joules. The distance between the centers of the bodies is r, measured in the appropriate units. L is the observable size of the Universe (about 12 billion light years or 1.1 E26 meters).

N_U is the total number of nucleons in the observable Universe and can be estimated from measured rotation and luminosity of galaxies. Astronomers calculate from stellar emissions that N_U is approximately 4E79 nucleons. As shown in Appendix G, we can calculate N_U using the measured value of Newton's G and obtain the value of 6.1E79 nucleons.

When L and r are measured in meters and the energy in joules, the force is expressed in newtons. There are no arbitrary dimensional constants in this gravity equation. All elements are independent variables or pure numbers.

Let's try out this new gravity equation.

Consider a brick having one-kilogram mass (conventional terms) resting on a weight scale on the surface of Earth. The Earth is composed of nucleons (electrons and protons—and neutrons, which we consider energetically nearly equal to a proton-electron pair). We count the number of nucleons (see Appendix C) and find N_1 of the earth is 3.6 E51 nucleons. We do the same for the brick and find N_2 of the brick is 6.0 E26 nucleons. The radius, r of Earth (nearly 4000 miles) is 6.37 E6 meters. Putting these values into the new gravity equation, we obtain a gravitational force of 6.5 newtons when just the visible matter of the Universe is considered, and 9.8 newtons when N_U is obtained from G. The classical Newtonian equation gives 9.8 newtons.

This new gravity equation will enable us to express Newton's gravitational constant, G, in terms of the parameters of the Universe and the speed of light. G is no longer an arbitrary constant. As we will see it tells us something important about the Universe.

Armed with this new insight we will be able to transform Einstein's "bending of light" equation and the escape velocity equation into simple ratios of nucleon numbers and distance. With the gravity code broken, the structure of our Universe becomes knowable and elegant beyond our wildest dreams.

Later, we will investigate the expansion of the Universe, and come to the amazing conclusion that the dynamics of expansion are independent of its matter. The quantity of "dark" matter as well as "bright" matter is irrelevant to the way in which our Universe expands. Our Universe is self-regulating.

Our next task is to derive the space-energy gravity equation.

CHAPTER 15

Deriving the Shape of Gravity

Let us begin our derivation of gravity with Einstein's discovery that the Universe is an unbounded, finite space, and astronomy's discovery that space has a uniform content of energy, represented by observable matter. The observed particles distributed in this space as stars, gas, and planets each have energy given by Einstein's mass-energy law. To repeat, we will consider the proton-electron pairs and neutrons as units of energy having most of the mass-energy of the observable Universe, and will call these units *nucleons*.

Our basic claim is; *each nucleon mass-energy is a condensation of some space-energy.*

The resultant depletion of space energy is inversely proportional to distance, and extends to the limits of space. The profile of the depletion of space-energy is shown in Figure 7. The gradient (slope of space-energy profile) of such a function is considered a virtual gravitational force. A measurable force will be the result of combining the effects of two or more depressions of space-energy associated with two or more nucleons. Figure 7 shows the space-energy decreasing as the particle is approached, from any radial direction. The distance scale is in arbitrary units. Space-energy, condensed to form the particle, is located arbitrarily at zero distance in this figure. The space-energy profile depicts the shape of space-energy after condensation. Gravity comes from the slope or gradient of this profile. The condensed energy that is the nucleon has a complicated energy distribution, packed into its extremely small volume, as is graphically suggested by the dip at the peak.

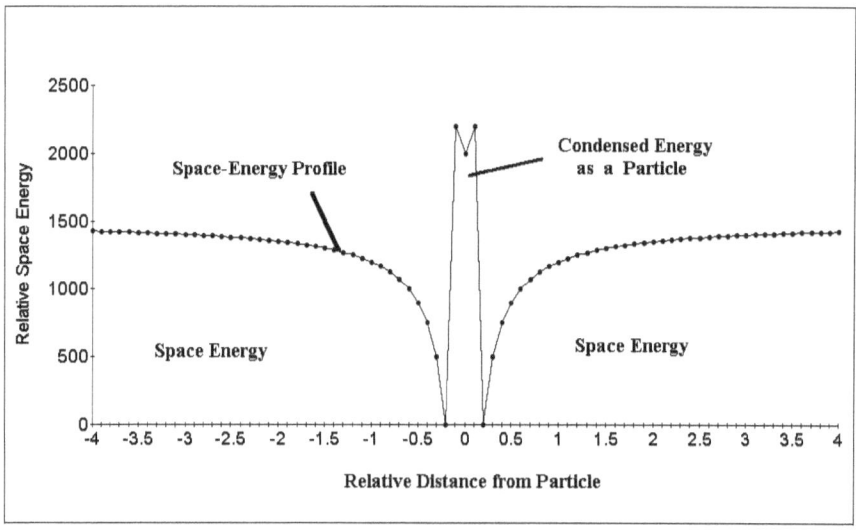

Figure 7. Condensation of some space energy becomes a particle. The result is a depletion of space-energy profiled here as energy in one dimension (vertical) with distance (horizontal).

Now as we enter some mathematical formalism we clarify what we are doing.

The assumptions are:

1. The Universe is a finite system (but unbounded), and limits the range of the addition or integration of energy.

2. Space energy exists,

3. Gravity is the space energy gradient,

4. Force is finite at any r (no infinities exist; the energy surface is continuous and the center is occupied by the finite mass-energy),

5. Energy is conserved: nucleon mass energy equals depression or depletion in space energy,

6. The mass-energy of mass M is given by Mc^2:

Terms: Given a nucleon mass-energy:

1. E_r is the depression in space-energy density at distance r

2. c is the speed of light, 3.00 E8 meters/second

3. L is the observable maximum dimension of space. L is approximately 1.1E26 meters.

4. N_U is the number of observable nucleons in the space defined by L and is approximately 6.1 E79.

5. E_u is the nucleon mass energy, 939 McV or 1.5 E-10 joules.

6. E_M is the mass energy for mass M.

7. Each mass M_i consists of N_i nucleons. The subscript i just labels any particular collection.

Let's start the derivation. At a distance r from a reference point, where the particle mass-energy will condense, the density of space-energy will change by an amount E_r. Based on Kepler's observations and verification of Newton's $1/r^2$ function, we may write the change in space energy density as inversely proportional to r.

$$E_r = K/r \qquad \text{joules/m}^3 \qquad (30)$$

K is, in mathematical terminology, a constant of proportionality. The variables are E_r and r. K has the same value for any E_r and r combination. We must evaluate K. We use conservation of energy, and the finiteness of space to find out what K is equal to. Since we know precisely how much energy is in the particle and we know the shape of the space-energy profile, we can find out how much energy was condensed at any distance r, from the particle. This information will give us K.

Going back to our sand-energy analogy, if I know how much sand is in the pile and I know the shape of the pit, I can determine how much sand was taken at any location. I can then get the absolute magnitude of the slope (gravity) of the pit at any location.

The energy that condensed from space-energy to form an electron or proton leaves a depletion in space-energy having the 1/r profile as discovered by Newton. This gives us our Equation 30 with the constant K. The space surrounding the particle can be divided into spatial shells having radius r and thickness Δr. Each shell has the space-energy density E_r. [Think of a huge set of nested bladder balls, each with thickness dr. The smallest ball (size r_c) is just a bit larger than the nucleon, and the largest ball is the size of the Universe (L). Each ball's wall is lacking a bit of its original energy density, used to make the nucleon.] The area of each shell is $4\pi r^2$ and the volume of each shell is $4\pi r^2\Delta r$. The collected energy at any r is simply the product of energy density times the volume of space at r. The amount of space-energy, partaking in the condensation, is obtained by adding all the shells of depleted space-energy surrounding the particle. This addition equals the mass-energy of the particle, $E_M = Mc^2$. We must do the addition from the particle (near zero distance) to the limit of observable space, L.

$$E_M = \left(E_r \, 4\pi \, r^2 \, \Delta r\right)_{SUMMED \; OVER \; ALL \; r} \quad \text{joules} \quad (31)$$

We use Equation 30 to substitute for E_r in Equation 31.

$$E_M = \left(K \, 4\pi \, r \, \Delta r\right)_{SUMMED \; OVER \; ALL \; r} \quad \text{joules} \quad (32)$$

Details of this summation are given in Appendix D.
The value of the summation gives us the mass-energy
of the particle,

$$E_M = Mc^2 = 2\pi \, K \, L^2 \qquad\qquad \text{joules} \quad (33)$$

which we solve for K.

$$K = \frac{E_M}{2\pi \, L^2} = \frac{Mc^2}{2\pi \, L^2} \qquad \text{joules/meter}^2 \quad (34)$$

We have been able to explicitly derive the form for the constant K
because energy is conserved and our Universe is finite, which limits the
addition of energy to the distance L. Using Equation 34 for K in
Equation 30, we may write for the energy density,

$$E_r = M \, c^2/2 \, \pi \, L^2 \, r \quad \text{joules/cubic meter} \quad (35)$$

We now have the energy density function, but we want the energy function itself to enable us to calculate the gravity force. Astronomy gives us the clue we need. When we observe space through a telescope, we see that the matter of space, in the form of stars, is collected in galaxies of stars having 100 billion stars or more in each galaxy. On the super-large scale of billions of light years distance, the distribution of galaxies is nearly uniform. This implies that in the early Universe, nucleons were uniformly condensed from space-energy. We also know that each nucleon has exactly the same amount of energy. From these observations, we may say that the space-energy, E_n associated with a nucleon is equal to the total energy E_U condensed from space energy to make all nucleons divided by the total number of nucleons, N_U,

$$E_n = E_U / N_U \qquad \text{joules} \qquad (36)$$

If the energy density used to form the nucleons is Er, then $E_U = Er\, V_U$, where V_U is the volume of our spherical Universe. Since $V_U = 4\pi L^3/3$,

$$E_n = E_r\, V_U / N_U = E_r\, 4\pi\, L^3 / 3\, N_U \text{ joules/nucleon} \quad (37)$$

But the energy density E_r is what we got from adding the shells of energy used to form the particle's mass-energy. Substituting in Equation 37 for Er, given by Equation 35, we get the energy profile for one nucleon,

Nucleon energy =
$$E_n = 2\, L\, M\, c^2 / 3\, N_U\, r \qquad \text{joules/nucleon} \quad (38)$$

This is the space-energy distribution for one nucleon. Although we define the nucleon as being electrically neutral, and do this by using the proton plus electron as a unit, the proton and electron have their own unique energy profiles. The proton energy profile may look something like that given in Figures 8 and 9.

We must multiply Equation 38 by N_1 to get the energy for N_1 nucleons, which is the number of nucleons in a body of mass M_1. The energy profile for N_1 nucleons is E_1 where,

Energy Profile:
$$E_1 = 2 \, L \, M_1 \, c^2 \, N_1 \, / \, 3 \, N_U \, r \text{ joules/nucleon (39)}$$

This tells us that space energy for any particle with mass M (or collection of particles of mass, M_1) has a spatial distribution depending on the size of the Universe and the total number of nucleons in it. Let's expand the mass M in terms of the nucleon mass-energy to simplify this energy equation. Using Einstein's mass-energy once again,

$$M_1 = E_1/c^2 = N_1 \, E_n \, /c^2 \qquad (40)$$

The mass-energy of mass M_1 is E_{M1}, which is just the total number of nucleons, N_1 in mass M_1, times the energy per nucleon, E_n. At this point, we specify that we are talking about a specific collection of nucleons, N_1, just for convenience of notation. The number N_1 may range from 1 for only one nucleon to the huge number of nucleons in a galaxy or more. We will call the space-energy profile E_1, to relate it to the collection N_1. The space-energy profile now becomes,

Space Energy Profile for Body N_1 =

$$E_1 = \frac{2}{3} \frac{N_1^2 \, L}{N_U \, r} E_n = \frac{2}{3} \frac{N_1 \, L}{N_U \, r} N_1 E_n \quad \text{joules (41)}$$

This result is amazing. The space-energy profile for any body having one or any number of nucleons depends not only on properties of that body, but depends also on properties of the entire Universe. The space-energy distribution of a single nucleon is illustrated in Figures 8 and 9. The way in which space-energy is shaped around each body depends on the size of the Universe and how much matter it has. This says that properties of space at any location are a function of the entire Universe. And notice, especially, that every term in the space-energy profile equation can be evaluated by independent measurements. There are no arbitrary terms. Einstein, in deriving his General Theory, sought ways to curve the geometry of space, not its energy. As a result he had to include Newton's constant of gravitation and missed the simple elegance of Equation 41. Now we must see how this space-energy profile determines the gravity force.

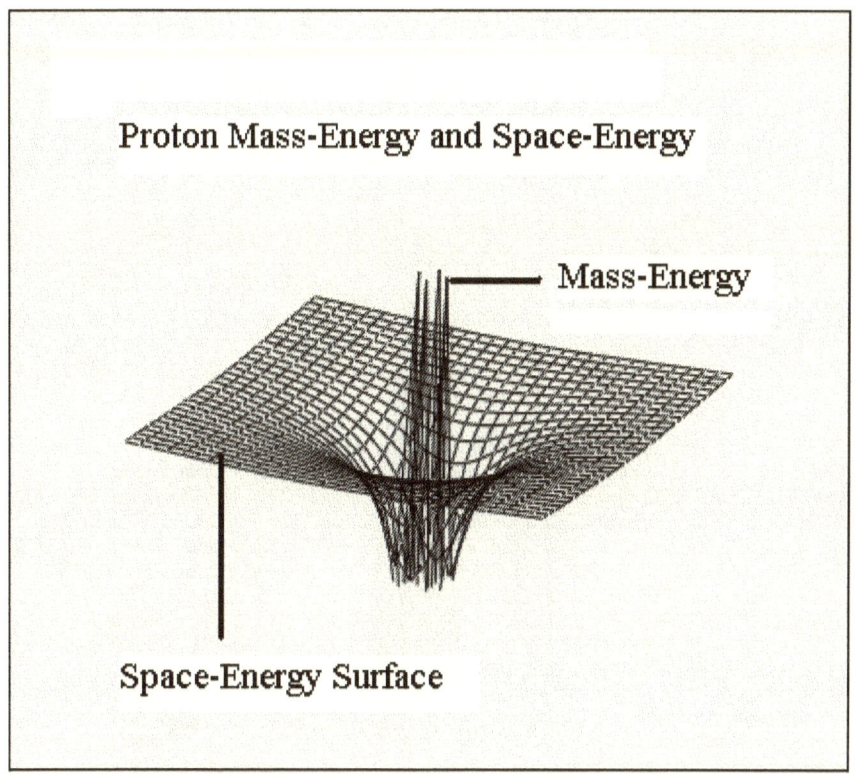

Figure 8. The mass-energy of a proton is confined to less than 1 E-14 meters, where it exists in resonance patterns, defined by the science of wave mechanics. We show the proton mass-energy here, in 3-D oblique view, with energy increasing toward the top of the page. The spike structure indicates sharp resonance states exist, and suggests that the energy gradients within the proton are extreme, as one would expect for the nuclear force. The space-energy profile gives the gravity force of the proton. The proton mass-energy is centered in its own space-energy hole.

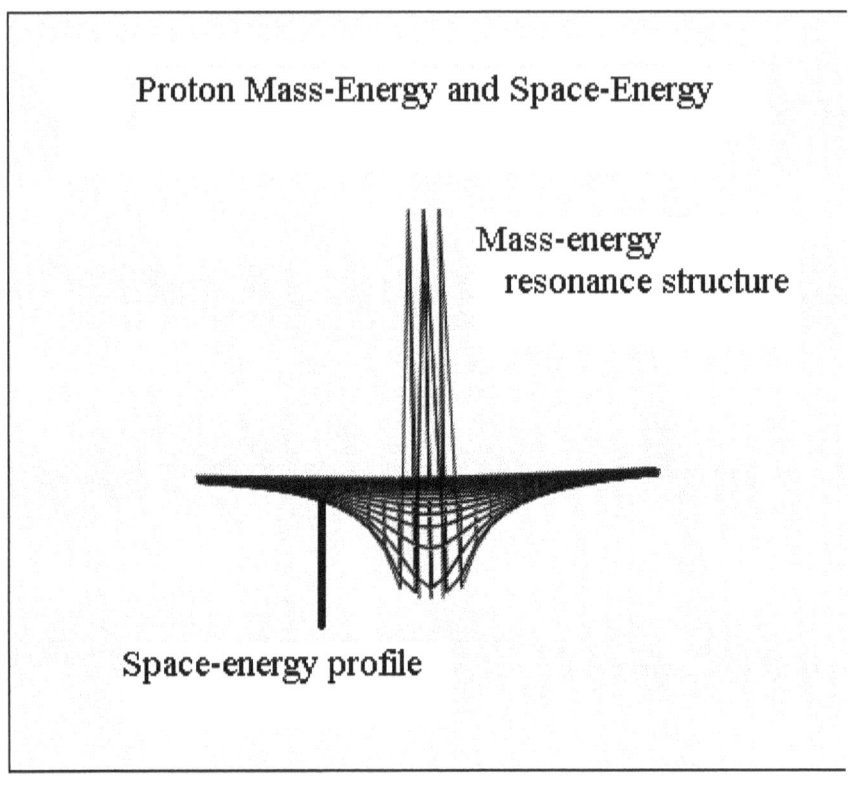

ET

Figure 9. Proton energy profile in edge view. See caption of Figure 8.

Consider now another collection of nucleons, N_2. Its space-energy profile will have the same form as collection N_1.

Space Energy Profile for Body N_2

$$E_2 = \frac{2}{3} \frac{N_2^2 \, L}{N_U \, r} E_n \qquad \text{joules} \qquad (42)$$

When the collections N_1 and N_2 are a distance r apart, the space-energy depletions overlap. Since space-energy must be conserved, the depletion of space-energy has a more complex symmetry as shown in Figure 10. The three-dimensional version of Equation 42 is shown in Figures 11 and 12, where one of the dimensions is energy, and the bodies are not the same size.

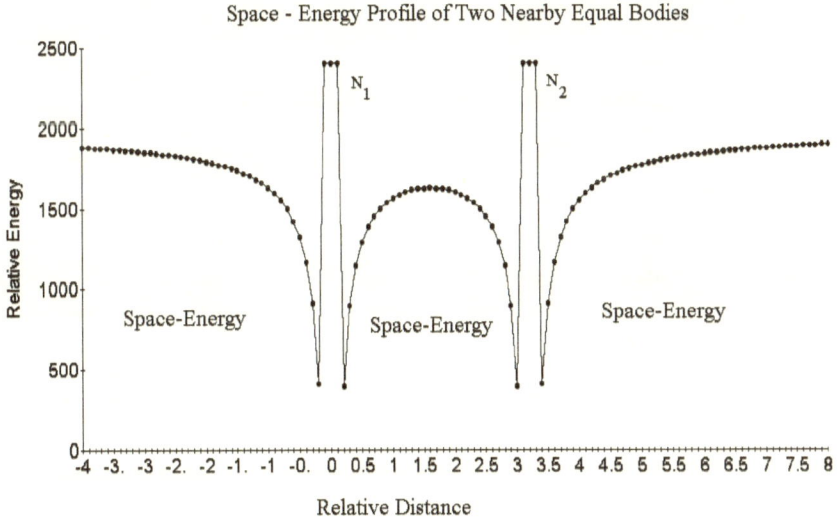

Figure 10. The space-energy profile for two equal collections of nucleons, separated by a distance of 3.2 units, shows reduced space-energy in the interim space. Both bodies compete for the available space energy.

This competition results in depressed space-energy in the interim space. If not held apart, the two bodies will move toward each other to establish space-energy symmetry for each nucleon.

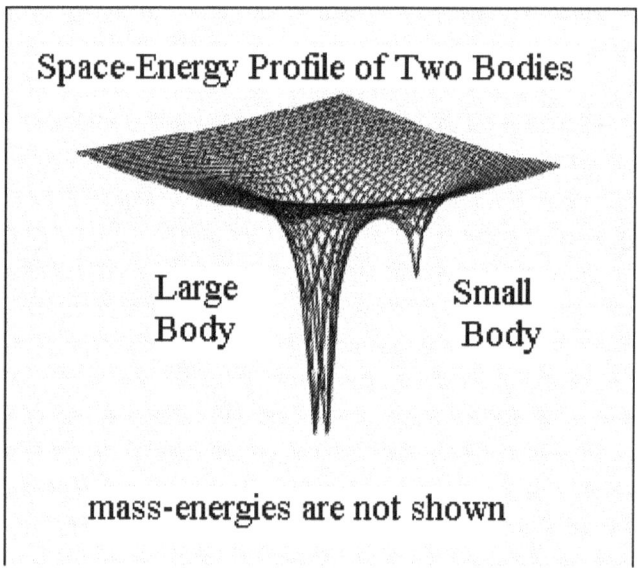

Figure 11. The space-energy profile of a large body and smaller body in oblique view, with the mass-energies left out for clarity. The gradients of the energy at all points on the energy surface are the gradient forces. The two bodies compete for space-energy, leaving weak repulsive gradient forces, as explained in the text. The result is stronger attractive forces pushing the bodies together. Gravity is the direct result of Least Action and Conservation of Energy.

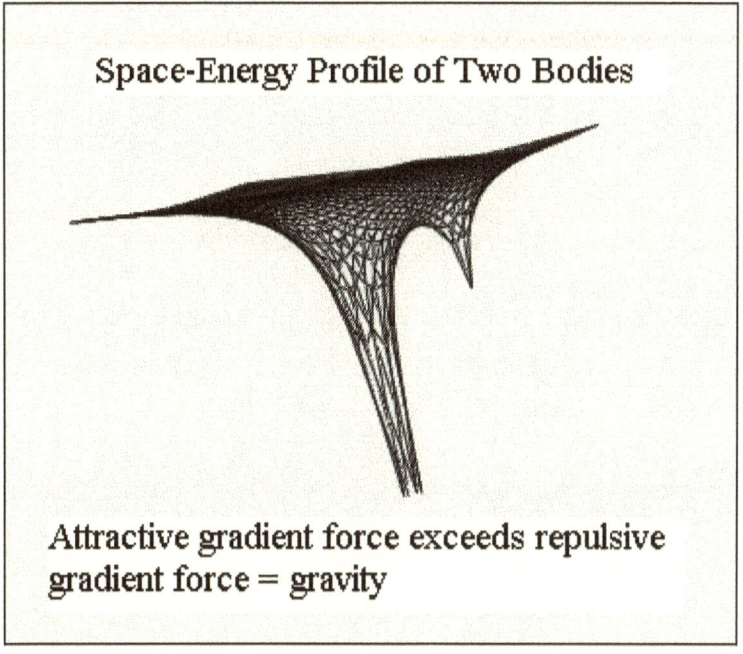

Etwo

Figure 12. Space-energy profile in the vicinity of large and small bodies in edge view. See caption of Figure 11.

*   *   *

Finding the Gravity Force

We now are ready to find the gravity force. Gravity is a force between nucleons or collections of nucleons. A brick is a collection of nucleons. Earth is a much larger collection of nucleons, and a star is larger yet. In the case of a brick, the nucleons that compose it attract each other by the gravity force and electric force. Compared to the electric force between a proton and an electron and even between atoms the gravity force is nearly infinitesimal. Without the electric force the brick would crumble to atomic dust by the gravity force of Earth. Gravity of the brick's particles is much too weak to hold the brick together. It is the electric force that gives shape and variety to all Earth's features such as mountains, trees, buildings, glass windows, ants, humans, bricks, virtually everything on Earth. The electric forces between molecules in a liquid are weak, and gravity determines its shape (lakes, seas, oceans). If the electric force vanished, the entire Earth would become a very small ball of dirty liquid. Gravity would hold the ball together. Gravity becomes a significant astronomical force for the cases where at least one of the collections of nucleons is large such a moon, planet or star. For convenience we will use *body* to mean collection of nucleons.

The space-energy profile for two bodies is a combination of E_1 and E_2. There is just so much space-energy allocated to form N_1 and N_2. As N_1 and N_2 approach each other, the depletion of space-energy must increase in the vicinity of N_1 and N_2 , with the greatest depletion between N_1 and N_2. Thus we see a greater depression of space-energy in the interim space. When Newton threw the apple to his neighbor's horse, the space-energy of the apple's nucleons was not symmetric about the apple because of the presence of Earth's space-energy. The result is the apple moving toward the center of Earth to establish a symmetrical space-energy distribution for the apple's nucleons. It is electric force that holds the apple together and it is electric force of the bits of soil material that prevents the apple's nucleons reaching Earth's center. Newton, the horse, and the apple are pinned to Earth's surface by the countering effects of gravity and electric forces.

This asymmetry of energy profile is the key to understanding the force of gravity between any two collections of nucleons. The profile gives us gravity and the clue of what must be done to achieve antigravity. We now calculate the gravity force.

CHAPTER 16

Gravity from Space-Energy

The tendency for a body to attain symmetry of space-energy (the same distribution in space-energy in all directions from the condensed energy) can be expressed as force. We find this force by examining the totality of gradients of space-energy profiles for any two or more bodies. The gravity force between two collections of nucleons is the sum of all the gradients of the combined space-energy profile, associated with all the nucleons in the two collections or bodies. The summation process can be greatly simplified by convolving, or putting together mathematically, the gradients for the two bodies. First, we take the gradient function of each body, separately. For example, a gradient is illustrated in Figure 13, as the slope of the energy profile, at a distance of 1.2 units.

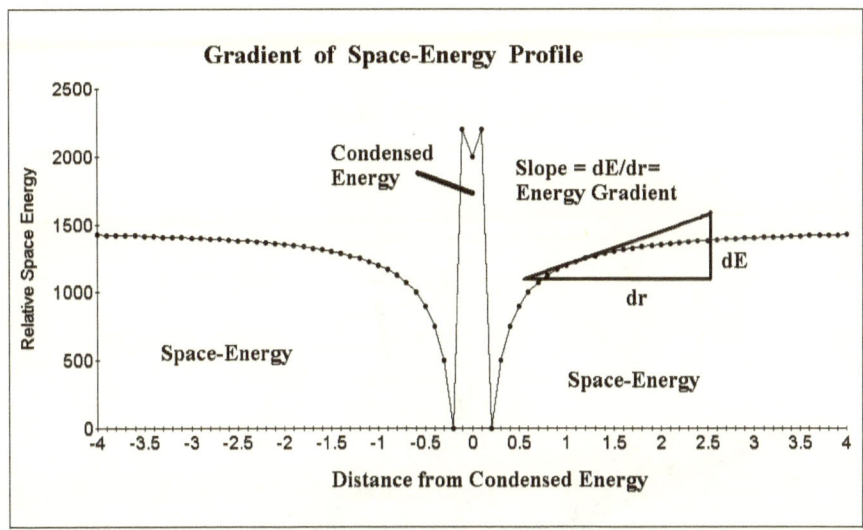

Figure 13. The energy profile of a body is shown in one space dimension (left to right), and the energy dimension increases toward the top of the page. The gradient force is the slope of the space-energy profile, at any point on the profile. As an example, we construct the slope at a distance of 1.2 units to the right of the body.

The energy shape for an extended body is shown in Figure 14. The mass-energy of a dense body (brick, moon, planet, star) looks like a tower of energy, and its space-energy depletion extends into all of observable space.

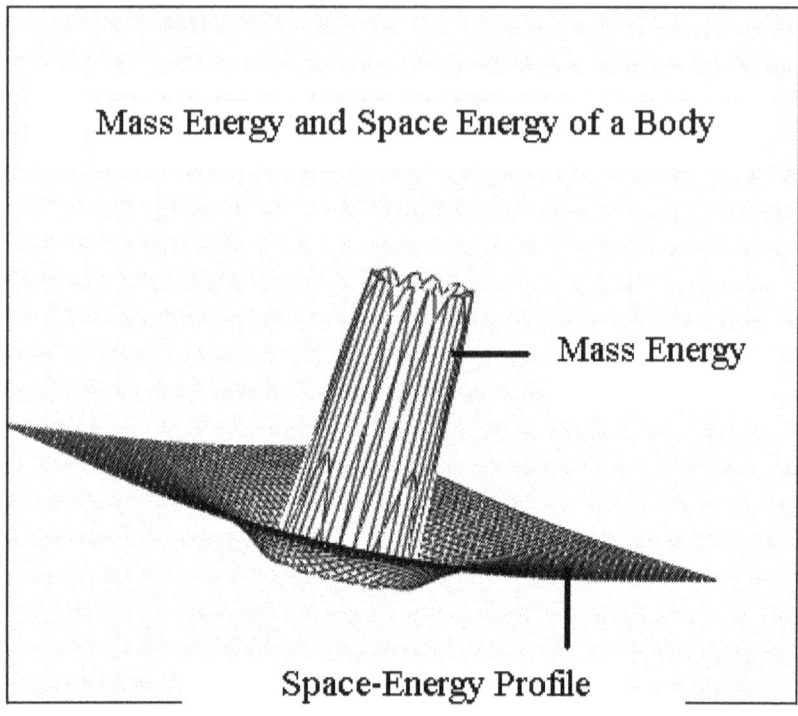

Figure 14. This wire-frame drawing shows the mass-energy core associated with an extended body such as a baseball, planet, or star. The space-energy profile defines the body's gravity. If this were the energy profile of Earth, we would be living at the energy surface, where the mass-energy meets the minimum space energy. We live in Earth's space-energy depression. We are in a space-energy hole. This is the energy a rocket ship must provide to reach Deep Space.

The gradient of the space-energy profile (gradient force) for any one body (collection of nucleons) at any distance, r, is considered a virtual force (a test body is needed to demonstrate force) associated with the body of particles N_1. That is, space exhibits a virtual force, F_1, for body

N_1 at distance r, and this force we call F_1, is given by the energy gradient (See Appendix F),

Gradient Force:

$$F_1 = dE_1 / dr = \frac{2N_1^2 E_n L}{3r^2 N_U} \quad \text{newtons} \quad (43)$$

We may now consider the energy density function for another collection of nucleons, N_2, located the distance r from N_1 as shown in Figure 10. It too generates a virtual force,

Gradient Force:

$$F_2 = dE_2 / dr = \frac{2N_2^2 E_n L}{3r^2 N_U} \quad \text{newtons} \quad (44)$$

The "combined " force is the **gravitational force** F_{12} of N_1 on N_2, and N_2 on N_1. We find the combined force by taking the square root of the product of the virtual gravity forces:

Gravity Force:

$$F_{12} = \left(F_1 F_2\right)^{1/2} = F_g = \frac{2N_1 N_2 E_n L}{3N_U r^2} \quad \text{newtons} \quad (45)$$

The gravity force is given by the product of N_1 and N_2, because N_1 times N_2 is the total number of interactions for N_1 nucleons and N_2 nucleons. Each nucleon attracts all other nucleons. Remember, E_n is the nucleon energy (1.5 E-10 joules), L is the maximum observable distance in the Universe (1.1 E26 meters) and N_U is the total observable nucleon count of the Universe (6.1 E79 nucleons).

N_1 and N_2 move toward each other to establish space-energy symmetry. We have found that this tendency to move may be expressed as a force, the gravity force. It is directed on a line connecting the centers of the two bodies to maintain maximum space-energy symmetry during the entire process.

The gravity force, as with the space energy depletion, is a function of the bodies and the entire Universe. And when expressed in terms of proper-ties of the Universe as was done above there are no arbitrary constants.

*   *   *

A new law of gravity:

The gravity force is the tendency of space energy to achieve symmetry and form one minimum for any combination of bodies.

Think of that. There is a property of space that regulates its very structure. Bodies in space are not independently acting with some kind of mysterious pulling force. Bodies are the stuff of space and each body exists in a space-energy well we have been calling a space-energy depletion. All space energy wells extend throughout space and they all interact. The tendency is for all energy wells to combine into one huge energy well in order to reach full symmetry. Space, by means of space-energy, is bringing all bodies together. We call this gravity and will explore its effect on our Universe below.

But first, let us try out the new gravity equation, # 45. A person (N_1), weighing about 154 pounds at Earth's surface, (see Appendix C) is in Earth orbit doing extravehicular maintenance on the International Space Station. All fitted out in a space suite and back pack the astronaut is composed of 7 E28 nucleons. With a mass of 130,000 kilograms or approximately 7.8 E31 nucleons, the space station and astronaut will be pushed toward each other by their space-energy gradients. They will seem to exert a mutual gravity force on each other. When the astronaut is at the far end of the solar panel, she may be approximately 50 meters (r) from the center of mass of the space station. Putting these values in Equation 45 we find that the apparent mutual attraction force is a meager 3 E-7 newtons (1 E-6 ounces). It is truly a micro-gravity environment.

The Shuttle and Space Station will have a mutual gravity of 1 E-4 newtons (about 4 E-4 ounces) at 50 meters separation. Shuttle maneuvers can pretty much ignore gravitational forces between it and the space station. What about landing on an asteroid?

An asteroid 10 kilometers in diameter is on a collision course with Earth. A small probe-rocket is quickly sent to the asteroid and put in orbit. Using the Kepler/Newton law of orbiting (Equation 16), the mass of the asteroid is measured and the nucleon content of the asteroid is calculated at 2 E42 nucleons. Astronauts are dispatched to land on the asteroid, bore a hole, plant nuclear charges, and fragment the asteroid to prevent Earth's destruction. Using Equation 45, where N_1 (asteroid) = 2 E42, N_2 (astronaut) = 5 E28, and r = 5000 meters, each astronaut will be pulled to the surface with a force of 0.56 newtons or only 2 ounces. Normal walking will be impossible and all tooling will have to be clamped to the surface. The Asteroid-Lander will have to move gently as in a docking maneuver. The asteroid must be treated like a giant space station. Since the asteroid's gravity is so weak, it may not take much explosive power to break it apart unless it is a continuous solid similar to the iron meteorites that have struck Earth before.

CHAPTER 17

Newton's Tension

What does the new gravity equation reveal about the structure of our Universe? And can it tell us how to diminish or eliminate the force of gravity? Gravity is a consequence of particles having formed from space-energy, probably during the beginnings of space itself. Is there a simple underlying structure of space that shows some kind of connectivity between what we observe and experience and the immensity of the Universe? Is the Universe a passive collection of particles or is it an active super-system vital to the ultimate destiny of its individual galaxies, stars and atoms?

In science, each result becomes a new tool for more learning, leading us on to greater discovery. What is, in fact, the impact of Equation 45? What does it reveal? To find out we will compare the new gravity equation and Newton's equation and try to figure out what Newton's constant is telling us. Once we know what G means, we can simplify some of the more interesting results of astronomical theory, and look for new physical insight into the workings of our Universe. Then, we will consider what it will take to achieve the neutralization of gravity.

We can now obtain the meaning of Newton's famous constant, G, by comparing the Newton equation with our space-energy equation. We convert M_1 and M_2 to their mass energies and express these in terms of the nucleon numbers and nucleon energy ($M_1 = N_1 E_n/c^2$ and $M_2 = N_2 E_n/c^2$),

Newton's equation becomes (in nucleon form):

$$F_g = \frac{GN_1 N_2 E_n^2}{c^4 r^2} \quad \text{newtons} \qquad (46)$$

Space-Energy Gravity Force:

$$F_g = \frac{2N_1 N_2 E_n L}{3N_U r^2} \quad \text{newtons} \qquad (47)$$

Solving these equations for G
(Equation 46 equals Equation 47) we have:

Newton's Constant:

$$G = \frac{2 c^4 L}{3 N_U E_n} \quad \text{meter}^4/\text{newton-sec}^4 \qquad (48)$$

And as we suspected, Newton's constant is a function of the properties of space. G is not a simple scaling constant, but what is it? Let's look at it mathematically upside down. If we take the mathematical inverse we have a function of force, $N_U E_n/L$. This force has the distinction of being the outcome of the structure of the Universe, namely, the total number of nucleons and the size of space.

Inverse G =

$$\frac{1}{G} = \frac{3}{2c^4} \frac{N_U E_n}{L} = \frac{3}{2c^4} F_U \qquad (49)$$

newtons / (unit speed)^4

Isn't it amazing? Here is a force function F_U that encompasses the entire Universe. What does it mean? Another word for force is *tension*. When we pull on a rope the pull-force in the rope is tension and the pull-force that breaks the rope is called its tensile strength. Is space in some kind of tension? We will explore this concept as we look at the dynamics of our Universe.

Another amazing thing about Equation 48 is that it can be used to find the total number of nucleons in the Universe. We can do a simple laboratory experiment to get a value for G, use our telescopes to measure the size of space L, measure the speed of light and the energy of a nucleon. We will then know how much matter is in all of space, whether we can see it or not. This calculation for N_U is given in Appendix G.

CHAPTER 18

Our Self-Regulating Universe

We have seen that gravity is a special property of space that defines the Universe. Albert Einstein, while developing his Theory of General Relativity, realized that gravity, as a universal concept, does not permit a static Universe. Space must be either expanding or contracting. If it is expanding due to some initial expansive condition, such as excess initial energy, it may eventually stop expanding and contract under the influence of its own gravity. But in the year 1912, Einstein, convinced by astronomers of that era that the Universe is static as if there is some kind of force counterbalancing gravity, inserted a term in his theory to include this unknown force. He called it the "cosmological constant" implying that it maintains a constant or static Universe. During the 1930s, when convincing evidence that the Universe is actually expanding, was obtained, Einstein, with gracious humility and subtle humor, called this extra term "my greatest mistake."

Knowledge of the expansion of our Universe was not well established until the middle of the 20th century. Before the 1920s our Universe was considered by astronomers to be composed mostly of stars and nebulae (fuzzy bright balls of something). The size of the Universe was not known. As telescopes got bigger, astronomers learned that many of the nebulae are isolated clusters of stars and that our solar system is located in an outer neighborhood of our home cluster we call the Milky Way.

Edwin Powell Hubble, while a graduate student and boxer at Chicago University, began to study optical spectra of isolated nebulae, which are island galaxies located outside of the Milky Way. Hubble, the greatest American astronomer, had just come back from serving on the American Front during World War I. He had studied law at Oxford, but found it too difficult to deal with clients. He loved astronomy and was lucky enough to work under George Hale at the new Mount Wilson Observatory. Using the 100-inch diameter telescope, Hubble studied extragalactic nebulae and found them to consist of billions of stars. The average distance between galaxies is 10 million light years. It takes 10 million years for light to travel between galaxies. Andromeda, our nearest neighbor galaxy, is only 1.5 million light years distant and can be seen without any optical aid on dark, clear nights. Like our Milky Way galaxy, Andromeda has a spiral shape, and is about 100,000 light years from end to end.

In order to understand how the Universe works astronomers must be able to determine astronomical distances. How does one measure the distance to anything in the Universe? Let's say you are on a hilltop at night and can see many thousands of city lights all around you. Some are very bright and many are dim. You may assume the bright lights are closer and all the dim lights are far away. But if a dim light is a very small light bulb it may be close by. When you blink your eyes left and right you notice some of the lights seem to move with respect to others. This is called the *parallax* effect. The nearer the light, the more is its apparent motion. By measuring the angle of shift and the distance between your eyes you can calculate the distance to some of the lights. Astronomers use this method and Earth's orbit to determine the distance to stars up to 300 light years away. Once the distance to any star is known, the effect of distance on its brightness may be eliminated. You may now calculate, for each of these stars, the *absolute brightness* , which is the brightness as if the star were at a specific distance. Of the thousands of lights you now know the distance to, you find that for most of them the absolute brightness depends on the star's color. The blue stars are brighter in the absolute-brightness sense. Now you can use color to find the distance to many more stars.

The big breakthrough in distance measurement for galaxies occurred when Henrietta Leavitt, an early twentieth century astronomer, discovered that a star, in the constellation Cephus varied rhythmically in brightness with a period of a few days. Measuring the period of many more such variable stars, now called Cepheid variables, she discovered that the brighter stars have longer periods of brightness variation. Since absolute brightness had become a measurement of distance, by use of the parallax and color methods, now the period of variation became a measurement of distance, extending distance measurement to the galaxies millions and billions of light years distant.

Hubble discovered these Cepheid variable stars in all nearby galaxies and was therefore able to measure the distance to each galaxy. Spectra showed that as a general rule, galaxies beyond a few hundred million light years were moving away from our galaxy. Using the Cepheid data and spectral data, Hubble discovered that the farther away a galaxy is, the faster it is going away from us. He measured this effect and found that the speed of recession increases nearly 15 kilometers/second for every million light years distance. This recession constant has become known as the *Hubble Term*. Edwin Hubble had discovered the galactic structure of the Universe and found that the Universe is expanding in an orderly, precise way. Later, other astronomers discovered that galaxies at distances of billions of light years are moving at a significant fraction of the speed of light. Hubble and others, believing that such speeds could not be possible, became critical of the spectral data and concluded in 1936, that the very distant galaxies were probably stationary. After WWII, Hubble joined the Mount Palomar team and helped build the 200-inch Hale telescope and was the first to use it. The immense improvement in data confirmed that the Universe is nearly uniform and expands at the rate given by the Hubbell Term to the very limits of observation. Galaxies 8 to 10 billion light years distant are receding from us at one third the speed of light. The limit of observable space is known as the *Hubble Length*. At the end of the 20th century, the Hubble Space Telescope had probed the depths of space and continues to make amazing discoveries.

In 1999 a team of astronomers lead by Wendy Freedman, an American, concluded after years of measurement that the Hubble Term equals 21.5 kilometers/second for each million light years distant from Earth. The uncertainty of this value is only 10%. This means that galaxies and quasars near the edge of the observational limit, which are going away from the Milky Way at 0.92 times the speed of light or 2.76 E5 kilometers/second are v/H = 2.76E5/21.5 = 12.8 E4 million light years = 12.8 billion light years away. The light from these distant galaxies has been on its way for 12.8 billion years.

Einstein showed with his General Theory of Relativity that the Universe can not be static. It can be expanding or contracting because gravity creates a non-equilibrium condition. Gravity will slow down an expanding Universe and will speed up a contracting Universe. With expansion, L, the Hubble length, must increase, meaning that G must be increasing. Since the size of our Universe is increasing with time, the tension of space is getting weaker, allowing the space energy gradient of each nucleon to increase. Therefore *the gravitational force is increasing with time*. It is as if the expansion of the Universe results in a more flimsy cosmic tension. Space loses its tension or stiffness as the Universe ages. If the space-energy fabric weakens, the energy depletions deepen. The Universe is in a race with itself. As it expands, its matter gains more attractive force. All matter tends to pull together with a stronger force.

Galaxies, caught up in the expansion of space, recede from one another as the Universe expands. The galaxies farthest from us move away from us at speeds approaching the speed of light and our Milky Way is also moving at near the speed of light as observed by astronomers living on distant galaxies. Of course the light from galaxies 12 billion light years away, has been in transit for 12 billion years, and started on the way to us near the beginning of the Universe. In one year, traveling near the speed of light, the Universe expands one light year or only one part in 12 billion parts. Attempts have been made to determine if G changes with time by observing the orbit of our moon with precise laser reflection tools, but it is very difficult to measure such a small relative change in G.

In the early and much smaller Universe, space exhibited a high level of stiffness; gravity, weak as it is now, was virtually ineffective. Matter-energy initially expanded unimpeded with expanding space-time. If this was the case, we would expect distant galaxies and quasars to have formed during a period when the gravity force was weaker than it is today.

Let us see if we can figure out if our expanding Universe will stop expanding. In 1924 Alexander Friedman, a Russian mathematician, showed that the fate of the Universe is a function of its total energy. The motion of the expansion gives us its kinetic energy and the force of gravity gives us its potential energy. The sum of these energies must be constant because total energy is always conserved. The kinetic energy per unit mass is simply $\frac{1}{2} v^2$ and the potential energy per unit mass is just $-GM_U/R$, where M_U is the mass of the Universe. The minus sign tells us that the gravity potential increases negatively with distance. This convention is necessary to retain the correct form when the energies are added. Summing these energies, we have:

KINETIC ENERGY + POTENTIAL ENERGY = CONSTANT

$$\tfrac{1}{2} v^2 - G (M_U/R) = - \tfrac{1}{2} k c^2 \qquad (50)$$

At a distance R from some arbitrary reference point, such as Earth, the unit mass is moving at speed v. The constant term includes the speed of light c to maintain the proper dimensions.

If the kinetic energy term is greater than the potential energy term, k will be less than zero (negative) and the Universe will expand for eternity. If, on the other hand, the potential energy is greater than the kinetic energy, k will be greater than zero (positive) and gravity will stop the expansion sometime in the future.

At the observable edge of the Universe, where R = L, v is the speed of light and we can rewrite Equation 50 to get:

Energy Condition of Universe:

$$\frac{1}{2}c^2 - \frac{GM_U}{L} = -\frac{1}{2}kc^2 \qquad (51)$$

Dividing by c^2 and multiplying by 2, Equation 51 becomes:

$$k = \frac{2GM_U}{c^2 L} - 1 \qquad (52)$$

Equation 52 becomes a test to see what our Universe will eventually do: expand unhindered forever (k is less than zero), stop expanding in finite time and start contracting (k is more than zero), or expand and come to a halt as time approaches infinity (k=0). The last case is of great interest to astrophysicists because it sets a lower bound on how much mass the Universe needs to eventually stop the expansion. When k = 0 the mass of Universe is called the *critical mass*. To keep the numbers easy to handle we will convert mass to its equivalent mass energy, equate the mass to density times volume, and calculate the *critical density* as nucleons per cubic meter of space.

Mass in terms of nucleon number:

$$M_U = \frac{N_U E_n}{c^2} \qquad (53)$$

For k = 0,

Critical Density $$D_C = \frac{N_U}{\frac{4}{3}\pi L^3} \qquad (54)$$

Putting together Equations, 52, 53, and 54, and inserting known values for the speed of light, the dimensions of space, Newton's gravitational constant, and energy of a nucleon,

Critical Density:

$$D_C = 3\,c^4\,/\,4\,\pi\,L^2\,G\,E_n =$$

$$9.9 \text{ nucleons/cubic meter} \quad (55)$$

Astronomers have estimated the density of space using data from galaxy rotation and luminosity, and come up with an upper limit of only 1.2 nucleon per cubic meter. What happened to the other 8.7 nucleons/m³? This is a serious problem for astronomy because most astronomers believe k must equal zero to satisfy certain aesthetic notions of symmetry. As a result they are looking for the missing matter, and because they can not see it they call it *dark matter*.

But the hunt for dark matter may be in vain. Astronomers may want it, but our Universe doesn't need it. Let's take another look at Equations 52 and 53. Combining these and using Equation 48 to replace G, we have the remarkable result that k is independent of how much matter space has.

$$k = \frac{4c^4 L}{3 N_U E_n}\, \frac{N_U E_n}{c^4 L} - 1 = \frac{1}{3} \quad (56)$$

Just think, k is a positive number, no matter how much matter the Universe has. The most surprising thing about this expression is that k is a pure number and is therefore independent of matter. This tells us that the course of expansion is also independent of matter, whether we can see it or not. Since G increases with expansion, the gravity force between galaxies also increases. Our Universe has a built-in braking mechanism, independent of its matter and energy content. With expansion of the Universe, space stiffness decreases, and the space-energy gradient associated with each nucleon increases, tending to slow the expansion. In any case we do not need to look for a lot of dark matter to get the Universe to stop expanding. It regulates itself, independent of how much matter-energy it has.

This means our Universe will slow down its expansion rate, come to a stop and start contracting. In some far distant eon, the Universe will use its gravity to crush itself in a gigantic compression. In its last stages of contraction, the temperature will soar, atoms will disintegrate, and our fundamental particles will be reabsorbed by the space-energy from whence they came. It is so because the space-energy force we call gravity is itself evolving, just as everything is in our wonderful Universe. The Universe is totally dynamic; everything is in a state of change.

Our Universe is truly amazing. It is of immense size and yet is finite but unbounded. It is expanding at the fantastic speed of light. The variety of structures, life forms, planets, moons, and stars is mind boggling. It has galaxies of many different shapes; some of them are in the act of collision. And to think that all of this vast size and differentiation is based on only space-energy and its condensed stable forms we call electron, proton, and photon.

CHAPTER 19

The Cosmic Ratio

The gravity force equation may be rewritten more symmetrically in terms of the mass-energies. Physicists like to write symmetrical equations because most of the Universe exhibits lots of symmetry. Equations of symmetry give more insight into physical phenomena. Multiplying numerator and denominator of Equation 45 by E_n, and expanding it, we have:

$$F_g = \frac{2}{3} \frac{N_1 E_n}{r} \frac{N_2 E_n}{r} \frac{L}{N_U E_n} \qquad (57)$$

Notice that the second and third terms on the right are ratios of mass-energy to separation distance. These two terms have units of force. We will call them F_1 and F_2. But the really interesting term is the fourth term, which is a force we will call F_U because it is composed of the size of the Universe L and the total energy $N_U E_n$ of the Universe. F_U is the universal energy gradient force. Gravitational force now has the appearance:

$$F_g = \frac{2}{3} \frac{F_1 F_2}{F_U} \qquad (58)$$

with

$$F_U = \frac{N_U E_n}{L} \qquad (59)$$

Looking at it this way, the gravity force is a product of local forces divided by a universal force (see Equation 59). F_U is a measure of the stiffness of space, and presently has the approximate value of 1.1 E44 newtons. It is a measure of the tension of space-energy. Gravity is a force of space-energy. It depends on the magnitude of space-energy depletions, which are complimentary functions of the matter-energy in each body and the distance between bodies. But the depletions themselves depend on the tension of space-energy. Any anti-gravity measure must deal with space-energy profiles, which includes this tension of the fabric of space.

Now, let's have some fun. We will establish a special dimensionless ratio of local and universal terms and see what happens to the equations of rocket escape speed as well as Einstein's deflection-of-light-by-matter derivation.

Once we have expressed Newton's G in terms of space-energy and dimension, some of the equations derived for motion simplify to only a ratio of numbers and dimensions. To see this, we define the "cosmic ratio" for the Universe,

Cosmic Ratio: $U_R \equiv L\,N_1\,/\,r_1\,N_U$ (60)

for a star or planet or any body with N_1 nucleons. A phenomenon is measured a distance r_1 from the center of the star or planet or body. Once again, L is the size of our observable Universe, and N_U is the number of nucleons in our Universe. Let's see what happens to escape speed and deflection of light by an astronomical body.

*   *   *

ESCAPE SPEED

Here's how rocket engineers calculate the escape speed or escape velocity of a rocket launched from a planet or moon having radius r_1. It is the minimum initial launch speed needed for the rocket to fully escape the gravitational pull of the planet and be able to reach deep space. It may be calculated by equating the kinetic energy of the rocket, to the change in gravitational energy, in going from the planet's surface to deep space.

Using the cosmic ratio, the equation for escape velocity is a trivial function of U_R. It is essentially the square root of U_R times the speed of light. The escape speed is a measure of the ratio of nucleons in the planet to that of the entire Universe. We can write the escape speed as a fraction of the speed of light using U_R to calculate the appropriate fraction. Check out Appendix E to see how this is done.

Escape Speed:

$$V_{ESCAPE} = \left(\frac{2GM_B}{r_1} \right)^{1/2}$$

$$V_{ESCAPE} = \left(\frac{4}{3} \frac{LN_X}{N_U r_1} \right)^{1/2} c = \left(\frac{4}{3} U_R \right)^{1/2} c \qquad (61)$$

Isn't this interesting? Escape speed is nothing more than a simple fraction of the speed of light.

Let's try this calculation for the planet Mars, which is composed of approximately $N_X = 3.8E50$ nucleons, and has a radius of 2.44E6 meters. Remember $L = 1.1$ E26 meters, $N_U = 6.1E79$ and $c = 3$ E8 meters/sec. The value of U_R for Mars becomes 2.2 E-10 and from Equation 55, we get nearly 5100 meters/sec or around 11,000 miles per hour, which is just about the launch speed needed, for the rocket to escape Mars and return to Earth.

When the escape speed equals the speed of light, not even light is fast enough to escape the pull of gravity. Light becomes trapped and the space body emits no light. It is a black hole. Setting $V_{escape} = c$, Equation 61 gives for the black hole condition:

Black Hole Condition:

$$r_1 = 4/3 \ (N_X/N_U) \ L \text{ meters} \qquad (62)$$

The escape horizon, r_1, is known as the Schwarzschild radius, after Karl Schwarzschild. If the sun were to gravitationally collapse from its 800,000 miles diameter to less than 1 mile in diameter, any light generated less than 1.4 miles from the center could not escape. However astrophysicists calculate that the sun does not have enough matter to pull itself into a black hole when it runs low of its matter-energy fuel billions of years from now. Instead it will become a red giant (huge star glowing red) and much later collapse into a very small, white hot star (white dwarf).

*   *   *

LIGHT DEFLECTION BY MATTER

Einstein explained that light is gravitationally deflected by the mass of a star, because the light follows the curvature of space, caused by the mass of the star. We now know that the photon's space-energy depletion interacts with the star's space-energy depletion to establish space-energy symmetry. Using the cosmic ratio, Einstein's equation for angular deflection of the light beam (measured in radians) simplifies directly to the Universal ratio:

Angle of Deflection: $\qquad \alpha = \dfrac{4}{3} U_R \text{ radians} \qquad (63)$

Astronomers, during a solar eclipse in 1919, verified his prediction. The value of U_R for our sun is 2.3 E-6, giving us 3.1 microradians as the angle of deflection for a light ray passing close to the sun. Notice the simplicity of Equation 63. The effects of gravity can be described by simple ratios of number and size. Today astronomers observe light from very distant galaxies, having been deflected by other galaxies on the path to Earth. This effect is called *galactic lensing* because the bending of the light has an effect as if the intermediate galaxy were a giant lens.

Even more interesting is the deflection produced by a black hole. Deflections of several degrees would be expected. This effect could be seen if the black hole were a companion to a star, so its apparent position would change as the star and black hole revolved around one another. If the black hole-star system's orbit were on edge to our view, the starlight would be considerably distorted during near coincidence.

CHAPTER 20

Beyond Gravity to Antigravity

We now know what causes gravity. It is the variation in space-energy associated with each condensation of energy we call a particle and the tendency of any space-energy depletion to obtain symmetry. The Universe is self-balanced by its space energy. Although changes in space-energy (which carry information) propagate no faster than the speed of light, the tension of space-energy itself may not have any particular speed limit. There is still much to learn about the workings of the Universe. Is there a way to alter the space energy gradients to achieve gravity shielding, or perhaps some kind of gravity repulsion? Perhaps there is. Read on…..

The search is on for anti-gravity. Governments are funding considerable research based on particle and quantum field theories. Individuals are delving into heavy mathematics looking for clues hidden in esoteric symbolism. Physicists who study extremely high-energy particles are building super detectors to attach to their matter-antimatter accelerators in hopes of discovering the next super-particle. Much of this activity has been reported in the popular press, and scientists generate book after book telling, telling…really not much. Because of our near sightedness, we only find what we are looking for. If you look in every nook and cranny of a mathematical equation, you are bound to find but a few nooks and crannies. It's going to take a big change in thinking to get to the next plateau of understanding. Can we achieve antigravity? First we have to ask what does antigravity mean? If we can figure this out, we can at least devise a plan for fruitful investigation.

Gravity, as we have seen, is the Universe's way of collecting all its matter (condensed energy and photons) over many eons into a gigantic super crunch. Each and every particle is sitting at the bottom of its space-energy depletion, which we call a *gravity well*. Particles compete for available space-energy. And since energy must be conserved exactly, the gravity wells lose their symmetry (see Figure 12), and the resultant asymmetric forces drive the particles together to form planets, stars, solar systems, and galaxies. Each body in the cosmos, no matter how big or small competes for space energy. The Universe is a collection of distorted gravity wells.

We Earthlings, like everything else, are collections of particles that are located at the center of our own gravity wells. Each one of us has a gravity well. My dog, Brook, has his own gravity well. Earth is big and dwells within a huge gravity well it is in the midst of. Brook's well and my well are pushing toward each other but the effect is extremely weak compared with our individual gravity wells in relation to Earth's gravity well. My center and Earth's center are being pushed toward each other. Earth is big, and I am small, so I do most of the moving. But Earth is hard (at least on land) and I don't go below the surface. The combined asymmetry of my well with Earth's well push on me and Earth with a force of 150 pounds. If a hole opens up in the floor, I fall toward Earth's center and Earth falls toward my center. When I and the bottom of the hole hit each other, I will hurt. Earth will come out OK. What would an antigravity machine do in a situation like this?

If this were a science fiction film the machine operator would yell out, "subspace energy warp!" and adjust the "graviton control," and I would float over the hole, saved from broken bone syndrome. The made-up words "subspace energy warp" and "graviton control" sound scientific but we do not know how to nullify gravity. Not yet. The machine would have to correct the space-energy asymmetry caused by the interaction of my gravity well and Earth's gravity well. It would have to increase depleted space energy so that my gravity well and Earth gravity well become symmetric. When my gravity well is uniform around me, I will not experience a gravity force. The International Space Station and its astronauts have symmetric gravity wells. It is achieved the hard way. With a lot of rocket fuel the space station sections and cosmonauts were brought up to orbit speed and once in orbit, the kinetic energy of orbiting restructures their gravity wells to near-perfect symmetry. The result is weightlessness.

Some scientists are looking at the possibility of being able to shield an object from Earth's gravity. Ordinary matter cannot shield against gravity because all matter carries with it, its own space-energy depletion. And as we have seen any two bodies compete for the available space-energy, leaving a region of enhanced depletion between the two bodies. The resulting energy gradients, the forces, are always directed to push the bodies together. My glass of water, sitting on the table, is being pushed toward Earth's center, because there is less space-energy in the space between my glass and the center of Earth.

To achieve the state of weightlessness known as levitation by magicians requires the addition of energy to the space located between the body of interest and Earth. If we knew how to do this the space-energy of the body would become symmetric and the body would appear weightless. How much energy does it take? It would be precisely equal to the escape energy which equals half the mass times the square of the escape speed given by Equation 61. When we make the necessary substitutions for mass as energy and use Equation 60 for the Universal Ratio we get:

Levitation Energy = $(2/3)$ $(N_B N_E E_n L / N_U R_E)$

Where N_B is the number of nucleons of the body, N_E is the number of nucleons of Earth, E_n is the nucleon energy, L is the size of the Universe, N_U is the number of nucleons in the Universe and R_E is the radius of Earth. Consider the levitation of a person weighing 150 pounds. Using the values in Appendices B, C and G we calculate that levitation requires the proper placement of 4.6E9 joules of energy. Think of that, levitation would be possible with only 4.9 billion joules of energy. This energy equals the output of a megawatt heater operating for an hour or a household kilowatt heater operating for 35 days. If electrical energy could be converted to stored space-energy and done efficiently it would cost approximately $150 at today's rate of 12 cents per kilowatt-hour to demonstrate levitation.

To achieve antigravity and repulsion, it will be necessary to reverse the net value of the space energy gradients. In effect, the net space-energy density located between any two bodies would have to be greater than the mean energy density of space. This is an energy problem. As bodies gravitate toward one another, they are moving to a region of lower space-energy. To stop or reverse this motion, energy must be supplied to the intervening space. There is no free lunch. Some other kind of energy configuration will be needed to provide the desired repulsive force. We must control space-energy and create a configuration of space-energy that would diminish or reverse the gravity force.

Another way of saying this is that we must make the space energy depletion of a given body symmetrical for it to be immune to the space energy gradients of any nearby bodies. To be weightless is to be symmetrical in space-energy.

Energy for Symmetry

Gravity is one of the four known forces that regulate and shape our Universe. The other forces are electric (also called electromagnetic force), weak nuclear, and strong nuclear. Electric and nuclear forces differ from gravity in many ways, but all are essential to have the kind of Universe we find ourselves part of. The three main differences are magnitude, polarity, and range. The nuclear force is some 40 magnitudes of ten (1 E40, or 10^{40}) greater than gravity and electric force exceeds gravity by 36 magnitudes of ten (1E36 or 10^{36}). Nuclear force, strong as it is, operates only in the small portion of space we call the atomic nucleus (about 1E-24 meters). Electric force, on the other hand, has the same dependence on distance as gravity, being effective over vast distances. But it is repulsive as well as attractive. Electrons repulse each other and protons repulse each other with force 36 magnitudes greater than gravity attracts. Protons attract electrons with a force equal in strength to electron-electron and proton-proton repulsion.

Electrically charged grids of metal, many millimeters from the electron beam of a TV picture tube, readily deflect electrons accelerated toward the tube face. Protons, emitted from the sun, are deflected by earth's magnetic field thousands of miles from Earth, suggesting the long-range effect of charge.

Einstein, in his search for a model to unify the fundamental forces of nature, looked for a connection between the gravity field and the electric field. He claimed no obvious connection exists, "…it may remain an open question, whether the theory of the electromagnetic field in conjunction with that of the gravitational field, furnishes a sufficient basis for the theory of matter or not." Let us revisit this issue but in terms of space-energy.

Living on Earth our bodies are not in a free-float situation. We feel our weight when space-energy forces us against rigid objects such as the floor. The floor is rigid because its atoms are bound together by the electric force. Our muscles develop in response to rigid bodies and gravity. Water is not rigid because molecules of water have a very small electric binding force, molecule to molecule. Let's compare these two forces, gravity and electric.

The gravity force is extremely weak compared to the electric force. If it were possible to couple but a very small part of the electric force or more precisely its energy gradient to the gravitational energy gradient, the gravity force would be dramatically changed. The electric force was defined and tested by Charles Augustin Coulomb in France, about the time of the American Revolution. Modern physics treats the electron as if it has no size, but has a property called charge. We must calculate the relative values of the gravity force and electric force, F_E to see what we are dealing with.

CHAPTER 21

Electric Force

Coulomb (and others) devised an equation for charge force, similar in structure to Newton's gravity equation. For a charge Q_A (say a collection of protons) and a charge Q_B (another collection of protons) separated by distance r, there will be a repulsive electric force.

Coulomb's Electric Force :

$$F_E = (1E\text{-}7)\ c^2\ (Q_A\ Q_B/r^2) \qquad (64)$$

What we know today is that charge comes in discrete units of electricity called coulombs, with each electron and each proton having exactly one unit of charge. This can become confusing. Remember that using the word "charge" is a convenient way of identifying a large force having attractive and repulsive features associated with the electron and proton. The magnitude of the charge q on each particle has been measured to be 1.6 E-19 coulombs. This unit of charge is valid when the constant in Coulomb's equation is equal to 1 E-7 times the square of the speed of light c measured in meters/second. The distance between charges r is measured in meters, and the force is expressed in newtons. Coulomb's equation works beautifully, but it does not explain charge.

Now that we have the equation for electric force we can obtain its magnitude relative to the gravity force in terms of the parameters of the Universe. Since the charge on a proton or electron has the discrete value of 1.6E-19 coulombs, the total charge of a collection of protons (or electrons) is just the number of protons, N_A times q. Thus $Q_A = qN_A$ and $Q_B = qN_B$. Coulomb's equation becomes:

$$F_E = (1E\text{-}7)\, c^2\, q^2\, (N_A\, N_B/r^2) \qquad (65)$$

This now has a look similar to our gravity equation, which is repeated here:

$$F_g = 2/3\, (E_n\, L\, /\, N_U)\, (N_1\, N_2\, /r_2) \qquad (45R)$$

Now, instead of N_1 and N_2 being neutral particles such as nucleons or neutrons, let us make them collections of protons. That is, $N_A = N_1$ and $N_B = N_2$. These packets of like charges will repel each other with a force of F_E and attract each other with a gravity force of F_g. What wins out? Let's take the ratio of forces to find out. The common terms cancel out leaving us with:

Force Ratio:

$$\frac{F_E}{F_g} = \left[\frac{3}{2} E - 7\right] c^2 q^2 \frac{N_U}{E_n L} =$$

$$1.6 \times 10^{36} = 1.6 E36 \tag{66}$$

We used c = 3 E8 m/s, q = 1.6 E-19 coulomb, N_U = 6.1E79 nucleons, E_n = 1.5 E-10 joules and L = 1.1 E26 meters. Think of that: the electric or charge force is over 36 orders of magnitude greater than the gravity force for any distance and any number of charges! How can this be? How can such a powerful force extend across the Universe? Can we tap into this electric force to counteract gravity?

How is it possible for the proton and electron to have such a strong force and yet, when combined into atoms of equal numbers of electrons and protons, the residual electric force a short distance away is almost zero? Atoms bind to atoms with electric energies in the 5 to 50 electron-volt range(one electron-volt of energy equals 1.6 E-19 joules) when the atomic spacing is less than 5 Angstroms (5 E-10 meters). This creates an atom-to-atom force much greater than the gravity force (the gravity force in a black hole is sufficient to reduce all matter to its basic energy components). The electric force is the primary force of the structure of matter. Our Universe is filled with the fantastic variety of things made possible with the electric force. The electric binding energy can take on over E20 (1×10^{20}) distinct configurations, giving us our hundred-odd basic elements and billions of chemical combinations. The electric force spices the Universe with amazing variety.

Physics simplifies the electric force by assigning a concept called *charge* to the electron and proton. No one knows what charge is, but somehow the electron and proton exhibit very powerful forces over long distances and with attractive and repulsive capability. Protons and electrons attract each other. Electrons push other electrons, and protons push other protons. Unlikes attract and likes repel. The concept of charge is handy, even if it obscures the actual physical process of proton and electron interaction. Since force is an energy gradient (change of energy with distance), the presence of the electric force implies that there is a special configuration of space-energy associated with the electron and proton. Up to now we have given these particles energy profiles that are monotonic (continually decreasing toward the particle's space-energy center). If, on the other hand, space-energy depletions are characterized by spatial waves, shells of space-energy surround each particle. This could give rise to very strong energy gradients as the energy shells of one particle interact with those of another. "Charge" may just be interactions of the proton and electron space-energy shells.

As an electron and proton approach each other, the space-energy shells form stable configurations given by the rules of wave mechanics. The resultant energy states of electron-proton configurations (atoms) are predicted by the wave equations of quantum mechanics. At distances greater than 1E-7 meters (1000 Angstroms) the energy shells of atoms are completely mutually nullified leaving only the space-energy gradient we call the gravity force. This leftover energy depletion is that which is equal to the mass-energy of the centralized energy concentrations.

Another fascinating property of the electric force is shielding. When an isolated piece of material is given an excess number of electrons, it will exert a force on another bit of charged material. The material can be an insulator or metal. It doesn't matter. When one of these charged bodies is surrounded by a metal shell, and the metal shell is connected to a large electrical sink, such as earth-ground, its electrical force does not extend beyond the shell. The metal shell shields the electric force.

These properties of the electric force suggest that "charge" is a special configuration of space-energy. Since this force can be repulsive as well as attractive, has a large magnitude, and long range - and can be shielded - the space-energy of electrons and protons must have the property of *phase*. When like particles are near each other, their space-energy shells are "in-phase" resulting in adding energies and strong repulsive force. When unlike particles approach each other, their space-energy shells are "out-of-phase" resulting in strong attractive gradients and forces. When two identical boats move in the same direction their waves combine both in-phase and out-of-phase somewhere behind the boats resulting in bigger waves in some places and near-flat water in other places.

The proton and electron, when in close proximity, have a nearly perfect phase mismatch, such that the electric force nearly vanishes, a short distance from a proton-electron pair. A pair of electrons has the combined space-energies in phase, giving greater energy slopes, greater force. Thus for electrons and protons taken separately, the force is attraction, until, when sufficiently close, phased energies overlap giving a stable atomic structure. But for electron-to-electron or proton-to-proton, the force is always repulsive. It is the phase relationships of the energy shells (oscillatory space-energies) that determine the direction of the force.

We must learn a lot more about the space-energy configurations of electrons and protons if we are to alter their phase relationships. It may then be possible to restructure the space-energy of a body, in a local sense, so that its space-energy gradient is made symmetrical. The result would be a reduction in the gravity force on that body. The guiding principle will be, as always, the conservation of energy. Roaring rocket engines, gorging on the released binding energy of billions of atoms each second, will be replaced by phase-altering engines that will change the local gravity. The energy equation will be the same. We live in a gravity-energy well and must pay the price for freedom. We'll just do it with less bluster. The key is to make the space-energy distribution about any body uniform. If we can remove the distortion, the body will become weightless.

It is always fun to go to a magic show and watch the magician levitate things that normally don't float in air. Magicians would use antigravity for all sorts of "magic" tricks. It would also be a wonderful tool for selectively reducing the weight of an object so that it could be moved around easily. Piano movers would love that kind of device. Hospitals could make their patients nearly weightless for certain kinds of therapy. The demand for antigravity devices would be enormous *if* one existed.

CHAPTER 22

New Directions

Gravity is the result of space-energy complying with the fundamental law of the Universe we call "conservation of energy." Gravity is the tendency of space-energy-depletions to merge into a single gigantic gravity well, as driven by space-energy gradients to achieve energy symmetry for any system of particles or bodies. Symmetry of space-energy represents the stable, lowest energy state of space-energy.

The photon, which reacts strongly with any electron-proton combination, is probably the key to breaking the electric-force code for the establishment of symmetry. We know a lot about how the photon interacts with matter, but we do not understand its basic structure. When we measure the photon as if it were some kind of a wave, it indeed acts like a wave with large spatial dimension. What is waving? We do not know. When we measure it as if were a particle, it acts like a particle of extremely small size. What is a particle? It is condensed energy. The photon is spread out and it is condensed. And so we say it exhibits the duality of wave and particle. This ambivalence just shows our ignorance.

The wave-particle duality of the photon reveals that the photon, too, has a space-energy component. But it is probably not a simple, monotonic, symmetric structure. The photon moves with a constant speed of 186,000 miles per second in a vacuum and somewhat slower in optical materials such as glass or water. Is it moving through space or is it a measure of the expansion of space? From the Einstein Theory of Relativity, we know that in the photon's internal clock system, time does not exist. It is anywhere on its path at the same instant. It can have phase properties connecting its source and its final absorption, even if these are separated by the vastness of the Universe.

What happens when the photon energy combines with matter? We say an atom has absorbed the photon energy, increasing the atom's energy. The atom's energy changes from one state to another in a discrete or quantized step. We do not have a detailed picture of the absorption and where or how the energy is stored. Quantum theory states that we can not measure the details of the absorption process since any measurement we might do will disturb the process. Even so we should attempt to theorize about the details and experiment on a large sampling, statistical basis. Then we shall know how matter is put together and perhaps how to achieve space-energy symmetry and zero gravity.

Finally, I offer a few comments on space-energy. It is a goal of physicists to adequately define any quantity or function used to describe physical phenomena. I have not explicitly defined space-energy, but I have assumed it to be a fundamental property of space. We experience its variations as gravity and electric forces but can not probe its depths. Typically, physicists consider energy to be a relative property of matter and not a quantity of itself. To say that space is imbued with energy, but that it is not in the form of particles, is contrary to modern theories of physics. To some, "space-energy" may sound like a revision of "æther theory" conceived prior to 1900 to explain the propagation of light. All experiments to find this æther failed. Hendrick A. Lorentz showed that because spatial relationships are relative, the presence of æther can not be detected by any physical experiment. If something can not be measured it does not exist and thus the existence of æther was considered meaningless. Is space-energy theory just revisiting æther theory? Does the photon require space-energy to propagate? Is space-energy density and the speed of light changing with the expansion of space? There are many questions left to answer. Physics of the 20th century broke new ground and greatly expanded our understanding of matter and the Universe. I am sure the 21st century will be even more fruitful.

Epilogue

The search for the "true" source of gravity is as least as old as the written word. Since the time of Newton, the degree that magic plays in notions of gravity has slowly diminished. But even in this era of Einstein people make little distinction between physical forces and spiritual influences. Most people know there are scientific explanations, even though they do not understand them. There are many rather wild ideas, ranging from "micro-quantum subspace" to "super strings." Search the Internet for "gravity", but look out, thousands of Links will be listed. The Web is a plethora of gravity sites, offering interesting ideas, confusion, and nonsense. Only a small fraction of it is good science. Search the libraries, and you will find that the space-energy concept of gravity almost appeared 40 years ago.

On their shoulders

The idea that the weakness of the gravitational force is related to the dimension of space was explored in the early 1960s. By equating the total gravitational potential energy of the Universe to its mass energy, the mid 20th century Relativist, David Sciama, had shown that $GM_U = Lc^2$. As above, G is the Newtonian gravitational constant, M_U is the mass of the observable Universe, L is the Hubble length and c is the velocity of light. If this relationship is used to replace G, the constant in Newton's gravitational law, we have:

$$F_g = M_1\, M_2\, L\, c^2\, /\, M_U\, r^2 \qquad (Ep1)$$

for the force between two masses, M_1 and M_2 at a centers' distance of r. This equation may be written in a simpler form, by replacing the mass quantities with their equivalent mass energies, and writing energy as a product of "nucleon" number and "nucleon energy." For example $M_1 = E_1/c^2 = N_1 E_n/c^2$ where N_1 is the number of nucleons and E_n is the nucleon mass energy. Although the word nucleon refers to proton and neutron, we assume a mass energy approximately equal to that of these particles, ignoring the effects of nuclear binding energy.

$$F_g = N_1\, N_2\, E_n\, L\, /\, N_U\, r^2 \qquad (Ep2)$$

Thus we have from earlier work our simple formulation for gravitational force, where the coefficient $E_n\, L/N_U$ is approximately 2.1 E-64 joule meters for the accepted values, E_n = 1.5 E-10 joules (939 MeV), L = 1.1 E26 meters (12 bly), and N_U = 6.1E79 nucleons.

This approach, although interesting, gave no definitive explanation for the magnitude of gravitation, or more importantly its source. But it did suggest that size of space is related to gravitational energy.

From there, we could speculate that the relationship may rest in a spatial energy density function for each nucleon or any particle having a measurable mass. The gravitational energy density function is not caused by the particle (as Einstein thought) but, rather, the gravity potential energy and particle mass-energy are *complimentary*. They exist together as a result of the condensation of space energy in the very early Universe.

If we assume the existence of energy associated with space-time as suggested by Ignazio Ciufolini & John Archibald Wheeler (well known modern cosmologists), we would have a structure capable of complimentary generation of particles and their variations in spatial energy or gradient forces. This concept was explored about 40 years ago by my professor of physics, Peter Bergmann who studied relativity at Princeton as a graduate student under Einstein.

The implications of this approach for cosmology suggest that the gravitational force was very weak in the early Universe, increased with expansion, and may itself provide the necessary deceleration to achieve a nearly flat or closed space. Physicists such as Alan Guth & James Steinhardt are diligently searching for more mass in the Universe and are proposing super-inflation theories to give the mechanics of the Universe a more ideal symmetry. It is almost as if expansion *must be* critically balanced by gravitation to obey some sort of grand design. One thing we can be sure of, the Universe is still full of surprises.

ABOUT THE AUTHOR

Sebastian Borrello studied physics at Syracuse University and was instructed in classical and quantum mechanics by Peter Bergman, who had been a graduate assistant of Albert Einstein. He spent the year 1958 as a geomagnetician stationed at Wilkes Base, Antarctica as part of the second International Geophysical Year. Working for Texas Instruments he adapted infrared scanning for medical diagnosis and worked with M. D. Anderson Cancer Institute in Houston to develop an infrared camera for early detection of breast cancer.

In1972 he derived the ultimate limit of photon detection based on the Uncertainty Principle, and he developed a theory of low-frequency noise based on fluctuations of crystal defects.

As a TI Fellow, he headed up a research team to combine infrared detection and signal processing in a single semiconductor chip. From 1979 to 1981, he was Chairman of the Detector Specialty Group of the National Infrared Information Symposium, sponsored by the US Navy. His team developed an all-silicon infrared detector, ideal for low cost commercial systems, for sensing toxic gases in the atmosphere, noninvasive blood sugar measurement, and IR cameras for fire fighters.

Sebastian has published many papers in the scientific literature, and has written on Photodetectors and Thermography for Kirk-Othmer *Encyclopedia of Chemical Technology*, fourth Edition, 1996 and 1998. He was born September 9, 1935, is married, and has two children.

APPENDIX

APPENDIX A:

Important Constants of the Universe

Speed of light, c = 2.9979 E8 meters/second

Proton mass-energy, E_p = 938.2 MeV = 1.50 E-10 joules

Electron mass-energy, E_e = 0.511 MeV = 8.18 E-14 joules

Least action, h = 6.625 E-34 joule seconds

Electron or proton charge, q = 1.602 E-19 coulombs

Newton's gravity constant, G = 6.67 E-11 newtons meter^2/kilogram^2

APPENDIX B:

Parameters of the Universe

Light year: distance light travels in one year = 5.9E12 miles = 9.46E15 meters.

Hubble length (observable limit of space), L = 12 to 13 billion light years = 7.1 to 7.7E22 miles = 1.14 to 1.23E26 meters.

Hubble Term (Expansion function of the Universe), H = 2.1 E4 meters/second, per million light years distance.

Observable galaxies = 1.5 E11

Stars per average galaxy = 2 E11

Nucleons in Earth = 3.57 E51

Nucleons in Sun = 1.2 E57

Total nucleons for a closed Universe, N_U = 1.0 E80

Total nucleons of Universe estimated from bright matter, N_U = 3.6E79
  Average density of our observed Universe = 1.5 nucleons/cubic meter

Total nucleons of Universe using the measured value of G, N_U = 6.1E79
Average density of Universe using measured G = 2.5 nucleons/cubic meter

Total nucleons for a relativistic closed Universe = 2.4E80
Average density of the relativistic closed Universe = 9.9 nucleons/cubic meter

(See Appendix G)

Appendix C:

The Measure of Things

Table C1. The dimensions of particles and bodies as size, mass, mass-energy and nucleons are listed for a variety of bodies. The number of nucleons in a body is given by its mass in kilograms divided by the mass of one nucleon, 1.67 E-27 Kg.

PARTICLE OR BODY	SIZE (DIAMETER)	MASS	MASS-ENERGY	NUMBER OF NUCLEONS
	Meters	Kilograms	Million electron volts ^	
Electron	*	9.11 E-31	0.511	
Proton	*	1.67 E-27	938.2	
Proton + electron	*	1.67 E-27	939	1
Water molecule H₂O	4 E-11	3.0 E-26	1.7 E4	18
Red blood cell	8 E-6	3E-14	1.7 E16	1.8 E13
Mustard seed	1 E-3	2 E-7	1.1 E23	1.2 E20
Golf ball	.043	.046	2.6 E28	2.8 E25
Human	1.7	70	3.9 E31	4.2 E28
Great Pyramid Of Cheops	115	5.2 E9	2.9 E39	3.1 E36
Moon	3.2 E6	7.35 E22	4.1 E52	4.4 E49
Earth	12.8 E6	5.97 E24	3.36 E54	3.57 E51
Jupiter	14.3 E7	1.90 E27	1.07 E57	1.14 E54
Sun	13.9 E8	2.00 E30	1.12 E60	1.20 E57
Milky Way Galaxy	1.0 E21	2 E41	1.1 E71	1.2 E68

* Less than 1 E-16 meters for mass-energy condensation of electron, proton, and proton + electron(in the neutron configuration).
^ One million electron volts (MeV) of energy = 1.6 E-13 joules.

Appendix D:

Summing the Energy Shells and Finding K

The integration of the energy shells, leading to Equation 34, is a spatial summing over the distance from the extremely small space of the condensed energy to the observable limit of space. The form of this integration is the same as finding the area of a circle. If a circle of radius r is divided into many concentric rings, the area will be the sum of the areas of all the rings.

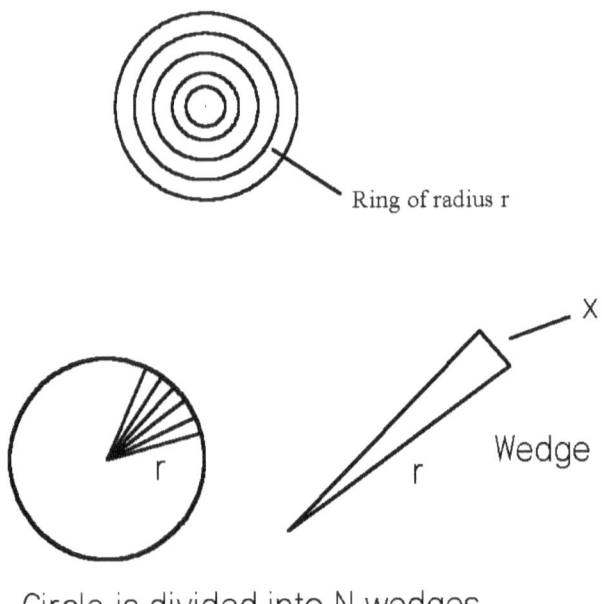

Ring of radius r

X

Wedge

Circle is divided into N wedges

Figure D1. Concentric Shells and wedges

The circumference, of a ring having radius r, is 2πr. The area of this annular ring is the circumference times the width w. When each annular ring has the same width the area of the circle included in r becomes,

$$A = 2\,\pi\,(r_1 + r_2 + r_3 + \ldots\ldots r)w \qquad\qquad D1$$

Mathematicians use a special symbol to denote a summation. The integral sign, having a stylistic S shape, is the instruction to add all elements with interval dimension dr. This form of addition is called *integration*.

Area of circle =

$$A = \int_1^r 2\pi r \, dr$$

(D2)

which has the well known value of πr^2 when r_1 is zero.

This result can be demonstrated in another way by dividing the circle into wedges and adding the areas of all wedges. Each wedge has an approximate area of rx/2 since the wedge is almost a triangle. But the wedge side x is equal to the circumference, $2\pi r/N$. The area of the wedge is $\pi r^2/N$. In the limit were the number of wedges, N, becomes infinite the area of the circle,

A = N (Area of wedge) = π r^2 (D3)

This is a demonstration that the integration given by Equation D2 is in fact, πr^2.

Finding K

The energy that condensed from space-energy to form an electron or proton leaves a depletion in space-energy having the 1/r profile as discovered by Newton. This gives us our Equation 30 with the constant K. The space surrounding the particle can be divided up into spatial shells having radius r and thickness dr. Each shell has the space-energy density E_r. the amount of space-energy partaking in the condensation is the integration over all the shells and equals the mass-energy of the particle,

E_M = Mc². We must integrate from the particle (near zero distance) to the limits of observable space, L.

We integrate (add) the change in space energy density, over all of space and set this equal to the mass energy E_M. For each distance r, there is a certain value of collected energy density E_r. The collected energy at any r is simply the product of energy density times the volume of space at r. At any r there is a spherical shell of area 4πr² and thickness dr, where dr is arbitrarily small. We add up all the shells of energy from the edge of the condensation energy, r_c to the observable limit of space, L, and equate this to the mass energy Mc². In mathematical shorthand we have,

$$E_M = Mc^2 = \int_c^L E_r 4\pi r^2 dr \qquad \text{(D4)}$$

The third term of this equation represents the addition of the energy density shells, concentric with the condensed energy dot. The term E_r depends on the distance, r as given by Equation 30 (E_r = K/r). Putting K/r in Equation D4 to replace E_r we have,

$$E_M = Mc^2 = 4\pi K \int_{r_c}^L r\, dr \qquad \text{joules (D5)}$$

The value of the summation from r_c to L is just $L^2/2$ (see the above discussion leading to Equation D2). The integral D5 yields, for the mass-energy of the particle,

$$E_M = M c^2 = 2 \pi K L^2 \qquad \text{joules} \qquad \text{(D6)}$$

and thus:

$$K = E_M / 2 \pi L^2 \qquad \text{joules/m}^2 \qquad \text{(D7)}$$

We have found K explicitly in terms of the mass-energy of a particle and the size of our Universe.

APPENDIX E:

Escape Speed from Planet X

When we launch our rocket ship from Earth, Moon or Planet X, it is necessary to supply sufficient energy from the burning rocket fuel to overcome the gravity force. Each astronomical body sits in its space-energy hole, and the gravity force at its surface is related to the depth of its space-energy hole. Planet X is our name for any astronomical body, be it planet, satellite, asteroid, comet or even a black hole.

It takes energy to climb a flight of stairs and a lot more energy to climb a mountain. When I climb, my muscles convert blood sugars to energy of motion at a fairly steady rate. My fitness and body physiology determine my maximum conversion rate. I can go up just so fast, but no matter what speed I move, the total energy is the same. I am moving partially up the space-energy hole of Earth, and the shape of this hole determines how much energy I will need to generate.

As an alternative, I can climb into the barrel of a cannon and let it shoot me to the top of the mountain. Now, the explosive gases must provide the entire energy needed to climb this part of Earth's space-energy hole, in the brief time I am in the barrel. If I could survive such crushing forces, I would leave the barrel with a specific speed. At sufficient launch speed, I could reach the mountaintop. Rocket ships are, in effect, self-contained cannons, expelling high-speed gases during the launch phase. When the speed at the end of the launch phase is sufficient to propel the rocket ship into deep space (far, relative to the size of Earth or Planet X), it is known as the *escape speed* for that particular astronomical body. Reaching escape speed means not falling back to Planet X. We can quickly calculate any escape speed, once we know the mass and size of Planet X. We will work it out.

The mass of our Deep-Space rocket ship, M_R, and its speed at the end of launch V_R, give us its kinetic energy, E_K.

Rocket Kinetic Energy =

$$E_K = \frac{M_R V_R^2}{2} \qquad \text{(E1)}$$

At any distance ,r, from the surface of Planet X, the energy needed to climb a little height, dr, is just the gravity force at r times dr. F times dr is the energy needed to go dr. The energy needed to climb the entire space-energy hole of Planet X is precisely the sum of all the F-times-dr products from the surface to deep space. When the rocket kinetic energy is equal to the climb-energy to deep space, the rocket speed, at the end of the launch phase, is the escape speed. We equate the kinetic energy with the space-hole-energy for a rocket of mass M_R escaping from a planet of mass M_X and radius r_1. The slim S sign indicates we are adding the increments of gravity force, F_g, times change in distance, dr.

Energy to escape =

$$\frac{M_R V_R^2}{2} = \int_1^L F_g \, dr \qquad \text{joules} \qquad \text{(E2)}$$

$$\frac{M_R V_R^2}{2} = \int_1^L G \frac{M_R M_X}{r^2} \, dr \qquad \text{joules (E3)}$$

We used Equation 4 to substitute for F_g in going from E2 to E3. The addition of increments extends from the surface of Planet X to somewhere in deep space, at a distance L. Using the rules of mathematics to carry out the addition of increments (called integration), we have:

$$\frac{M_R V_R^2}{2} = G \frac{M_R M_X}{r_1} \qquad \text{joules} \qquad \text{(E4)}$$

$$V_R = \left(\frac{2 G M_X}{r_1} \right)^{1/2} \qquad \text{meters/second} \qquad \text{(E5)}$$

The cosmic significance of escape speed becomes clear when we convert Planet X's mass to mass-energy, and use Equation 48 for G. The escape speed becomes a simple fraction of the speed of light.

Escape Speed:

$$V_R = \left(\frac{4}{3} \frac{L \, N_X}{N_U \, r_1} \right)^{1/2} c \qquad \text{meters/second (E6)}$$

We may put this in terms of the Cosmic Ratio as defined by Equation 60,

Escape Speed:

$$V_R = \left(\frac{4}{3} U_R \right)^{1/2} c \qquad \text{meters/second (E7)}$$

The escape speed may be calculated for any Planet X, once we know how many nucleons, N_X, it has and what its size, r_1, is. For example, when the ascent stage of the Lunar Excursion Module (LM) of the Apollo Moon Mission, blasted off Moon's surface, it achieved nearly the Moon's escape speed to reach the Command Module. For Moon we have, from Appendix C, $N_X = N_{MOON} = 4.4E49$ nucleons and $r_1 = 1.6\,E6$ meters. From Appendix B, $L = 1.1\,E26$ meters and $N_U = 6.1\,E79$ nucleons. Using Equation E6, the LM ascent stage escaped Moon at a speed of nearly 2150 meters per second (equals 1.3 miles per second or 4800 miles per hour).

The reverse of escaping is landing. Should the retrorockets fail to fire, the Lunar Lander would hit the surface at 4,800 miles per hour. Reliability is essential! Earth entry is a bit different. Earth's atmosphere is used as a brake, absorbing nearly all the energy gained during the plunge into Earth's space-energy gravity well. The heat shield must stay intact to prevent overheating and vaporization of the spacecraft.

Parking a rocket in orbit, before deep space injection, does not affect the energy requirements, just as pausing on the mountain trail, does not affect the total energy needed to reach the top.

APPENDIX F:

The Gradient Force

We say that the gravity force for a body is the gradient of the space energy for that body. The idea of gradient is taken from everyday experience. Consider a road that is not level. There may be a sign at the top warning "steep grade, use low gear." The word *gradient* is a descriptive form of grade. Sometimes the signs give the value of the grade as a percent, such as "warning 6% grade, use low gear." Now you know that for every 100 feet of road you go down 6 feet. A stairway, on the other hand, displays a discrete measure of grade in the form of steps. Typically, a 12-inch step forward is compounded with an 8-inch step up to give a gradient of 8/12 or 66.7 %.

In the case of a road, both length and drop are measured in feet or meters. The grade or gradient is the ratio of the drop to the road distance. The dimensions divide out leaving us with a numerical ratio or percent. In the case of a road, the force pulling your can down the hill is proportional to the grade. This force is a component of the gravity force, and has a direction parallel to the road.

The idea of gradient may be used when the vertical distance is replaced, conceptually, with energy and the horizontal is still a unit of distance in feet or meters. The structure of gradient is energy step divided by distance. If the energy step is 3 joules and the displacement is 4 meters, the gradient is 3 joules/ 4 meters or 0.75 joules per meter. The unit of joule per meter is conceptually force with its unit value known as a *newton*. Thus the gradient in this example becomes a force of 0.75 newtons. The only spatial direction we have is horizontal, which means the force has a horizontal direction.

The gradient steps can, in principle, be made arbitrarily small and no meaning is lost. In a practical sense the steps vanish but the idea and structure of gradient is unchanged. As the steps become infinitesimal, the mathematical rules of differential calculus may be used to achieve what are known as analytical solutions. Physicists strive to achieve clean, analytical equations to make it easier to understand what is going on. Equation 43 is an example of differential analysis.

It is obtained by differentiating Equation 41 with respect to the radial distance. Thus the force F1 for body N1 is:

$$F_1 = dE_1/dr = (2N^2_1 \ L \ E_n/3 \ N_U)1/r^2 \qquad \text{(F1)}$$

Formally there should be a minus sign to indicate the force is central or points to the center of the body

APPENDIX G:

Weighing the Universe

How much matter is in our Universe? How many nucleons is the Universe composed of? One way to find out is to count the nucleons in a star, count the stars in a typical galaxy, and count the galaxies. Astronomers have done this. Our sun is an average star with 1.2E57 nucleons and our galaxy is a typical galaxy with 200 billion stars. And recently the Hubble Space Telescope has probed deep space and finds approximately 150 billion galaxies. Putting these numbers together the Universe consists of N_L nucleons in the form of luminous matter.

Luminous Matter:

N_L = 1.2E57 x 2E11 x 1.5E11 = 3.6 E79 nucleons.

If we estimate that 10% of the Universe is tied up in planets, cold debris and galactic black holes, the total number of nucleons would be N_U = 4 E79.

Another and more elegant way to find N_U is to use Equation 48 which is repeated here.

Newton's Constant:

$$G = 2\ c^4\ L\ /\ 3\ N_U\ E_n \qquad\qquad (48R)$$

Each element of this equation is independent, allowing us to calculate any one element if the others are known. Scientists have measured the speed of light c in a variety of ways and get almost exactly 3.00 E8 meters/second. Calibrated data from the Hubble Space Telescope confirms the edge of observable space is some 12 billion light years distant (1.1 E 26 meters). Particle physicists and nuclear scientists have established that a nucleon has 1.5 E-10 joules of mass-energy. We are left with G and N_U. If we could see all of matter and count it, we could calculate G. However it is much easier to accurately measure G, as we have seen and use its value to find how much matter is in our Universe. Rewriting Equation 48,

Total Matter of our Universe:

$$N_U = 2\ c^4\ L\ /\ 3\ G\ E_n \qquad\qquad (G1)$$

Putting in the numbers we find, N_U = 6.1 E79 nucleons. Finally, with the help of Galileo, Kepler, Newton, Einstein and hundreds of other scientists we have measured the totality of space. Or have we? Relativists say this simple approach does not properly take into account the curvature of space when considered on such a grand scale. Their curvature correction increases their expectation of N_U by a factor of 1.5π, giving a value of 2.4E80 nucleons. If they are right, astronomers are seeing only 1/6 of matter in the form of luminous material. Where is the missing matter? Why is most of the matter of space dark? Our Universe is too young to have most of it in the form of dark matter. Whatever is the case about dark matter, Equation G1 gives us consistency and closure for the measurable parameters of space. Should we ask for much more than that?

BIBLIOGRAPHY

CONCEPTUAL AND DESCRIPTIVE

1. Motz, Lloyd and Jefferson Hane Weaver, *The Story of Physics*, Avon Books, New York, 1989.

2. Park, David, *The How and the Why, An Essay in the Origins and Development of Physical Theory*, Princeton University Press, Princeton, N.J., 1988.

3. Freedman, Wendy L., *The Expansion Rate and Science of the Universe*, Scientific American, Nov. 1992.

4. Einstein, Albert, *Sidelights on Relativity*, Dover Publications, 1983, lectures at the University of Leyden in 1920 "Ether and the Theory of Relativity", and the Prussian Academy of Sciences in 1921, "Geometry and Experience."

CHALLENGING BUT NOT COMPLICATED

5. Rothman, Milton A., *Discovering the Natural Laws*, Dover Publications, Inc. New York, 1989.

6. Taylor, Edwin F. and John Archibald Wheeler, *Spacetime Physics, Introduction to Special Relativity*, W.H. Freeman and Company, New York, 1992.

7. Sciama, D. W., *The Physical Foundations of General Relativity*. Doubleday & Company, NewYork, 1969

8. Dicke, R. H., "Gravitation - An Enigma" Lecture, American Scientist 47, 25 (1959).

9. Harrison, Edward R., *Cosmology, the Science of the Universe*, Cambridge University Press, N.Y., 1981.

10. Guth, A. L. & Steinhardt, P. J., *The Inflationary Universe*, Scientific American, Nov. 1992.

11. Silk, Joseph, *The Big Bang*, W. H. Freeman & Company, San Francisco, 1980

ADVANCED

12. C. Brans & R. H. Dicke, Physical Review. 124, #3, 1961, p925.

13. Ignazio Ciufolini & J. A. Wheeler, *Gravitation and Inertia*, Princeton University Press, 1995.

14. Peter G. Bergmann, Introduction to the *Theory of Relativity*, Prentice-Hall, New York, 1955.

15. H.A. Lorentz, A. Einstein, H. Minkowski and H. Weyl, *The Principle of Relativity*, Dover Publications, Inc. New York, 1952. Translation of 1923 German edition.

INDEX

0-595-20969-6

www.ingramcontent.com/pod-product-compliance
Lightning Source LLC
Chambersburg PA
CBHW030944180526
45163CB00002B/691